Thermo-Acoustics of Nanofluids and Transfer Processes

This book explains the physical principles and theoretical basis of acoustics of nanofluids with mathematical rigor, focusing on concepts and points of view that have proven effective in applications such as heat transfer, petroleum science and technology, wastewater treatment, food processing, and hydrogen production. It provides a rigorous framework to aid readers in building innovative nanofluid-based devices, covering essential topics such as the study and measurement of thermophysical characteristics, convection, and heat transfer equipment performance.

Features:

- Focuses on the basics of nanofluids, shedding light on the thermo-acoustic behavior.
- Discusses the specific needs of a nanofluid for a process in terms of both heat and mass transfer.
- Discusses the process transfer of nanofluids with reference to thermo-acoustics.
- Discusses the numerical and experimental investigations of nanofluids used in various fields of industrial applications.
- Reviews fundamentals and applied aspects of acoustic cavitation.

This book is aimed at graduate students and researchers in fluid dynamics, nanotechnology, and chemical and mechanical engineering.

Emerging Materials and Technologies

Series Editor: Boris I. Kharissov

The *Emerging Materials and Technologies* series is devoted to highlighting publications centered on emerging advanced materials and novel technologies. Attention is paid to those newly discovered or applied materials with potential to solve pressing societal problems and improve quality of life, corresponding to environmental protection, medicine, communications, energy, transportation, advanced manufacturing, and related areas.

The series takes into account that, under present strong demands for energy, material, and cost savings, as well as heavy contamination problems and worldwide pandemic conditions, the area of emerging materials and related scalable technologies is a highly interdisciplinary field, with the need for researchers, professionals, and academics across the spectrum of engineering and technological disciplines. The main objective of this book series is to attract more attention to these materials and technologies and invite conversation among the international R&D community.

2D Materials-Based Sensors: Technology and Applications
Vinod Kumar Khanna

Metal Organic Framework Derived Materials: Design Strategies and Applications
Gomathi Nageswaran, Varsha M V, Arun Kumar Rajasekaran, and M Shashank Rao

Hydrogen Production, Storage, and Utilization: Technologies and Applications
Abbas Tcharkhtchi, Hamidreza Vanaei, Albert Lucas, and Sedigheh Farzaneh

MXenes for Energy Storage Applications: Emerging Characteristics, Compositions, and Synthesis Methods
Muhammad Rafique, M. Bilal Tahir, and Saira Anwar

Multi-scale and Multifunctional Coatings and Interfaces for Tribological Contacts
Ajit Behera, Kuldeep K Saxena, Dipen Kumar Rajak, and Shankar Sehgal

Nanotechnology in Green Energy Generation
Ahmed Thabet Mohamed

Thermo-Acoustics of Nanofluids and Transfer Processes
Shriram S. Sonawane and Manjakuppam Malika

Hybrid Nanofluids: Heat and Mass Transfer Processes
Shriram S. Sonawane and Manjakuppam Malika

For more information about this series, please visit: www.routledge.com/Emerging-Materials-and-Technologies/book-series/CRCEMT

Thermo-Acoustics of Nanofluids and Transfer Processes

Shriram S. Sonawane and Manjakuppam Malika

CRC Press
Taylor & Francis Group
Boca Raton London New York

CRC Press is an imprint of the
Taylor & Francis Group, an **informa** business

Designed cover image: shutterstock

First edition published 2025
by CRC Press
2385 NW Executive Center Drive, Suite 320, Boca Raton FL 33431

and by CRC Press
4 Park Square, Milton Park, Abingdon, Oxon, OX14 4RN

CRC Press is an imprint of Taylor & Francis Group, LLC

© 2025 Shriram S. Sonawane and Manjakuppam Malika

Reasonable efforts have been made to publish reliable data and information, but the author and publisher cannot assume responsibility for the validity of all materials or the consequences of their use. The authors and publishers have attempted to trace the copyright holders of all material reproduced in this publication and apologize to copyright holders if permission to publish in this form has not been obtained. If any copyright material has not been acknowledged please write and let us know so we may rectify in any future reprint.

Except as permitted under U.S. Copyright Law, no part of this book may be reprinted, reproduced, transmitted, or utilized in any form by any electronic, mechanical, or other means, now known or hereafter invented, including photocopying, microfilming, and recording, or in any information storage or retrieval system, without written permission from the publishers.

For permission to photocopy or use material electronically from this work, access www.copyright. com or contact the Copyright Clearance Center, Inc. (CCC), 222 Rosewood Drive, Danvers, MA 01923, 978-750-8400. For works that are not available on CCC please contact mpkbookspermissions@ tandf.co.uk

Trademark notice: Product or corporate names may be trademarks or registered trademarks and are used only for identification and explanation without intent to infringe.

ISBN: 9781032622910 (hbk)
ISBN: 9781032977003 (pbk)
ISBN: 9781003594949 (ebk)

DOI: 10.1201/9781003594949

Typeset in Times
by codeMantra

*I dedicate this book to my loving family – my mother
"AAI," my father "DADA," my most loving wife Rina, my
gorgeous daughter "Sanyuja," and my son "Shaurya."
Your unwavering support and encouragement have been
my guiding light throughout this incredible journey.*

Shriram S. Sonawane

*This work is dedicated to my parents, Mrs. Nirmala and
Mr. Gopal, and my sisters, Mrs. Monica and Mrs. Harika,
whose unwavering support and encouragement
fostered my passion for science and technology.*

*I also extend my heartfelt gratitude to my beloved husband, Mr. Rahul
Mudliar, and my daughter, Ms. Prisha Mudliar, for their patience
and constant support throughout this journey. Special thanks to
my in-laws, Mrs. Sujatha Mudliar and Mr. Narendra Mudliar, for
their understanding and help during the course of this project.*

Manjakuppan Malika

Contents

Foreword

The convergence of nanotechnology and thermo-acoustic processes has facilitated innovative methodologies for enhancing heat and mass transfer activities across several industrial applications. Engineers and scientists are increasingly focusing on the research on nanofluids and their thermo-acoustic properties to improve energy efficiency contributing to sustainability. This book, *Thermo-Acoustics of Nanofluids and Transfer Processes*, is a pertinent and thorough treatise on this dynamic and swiftly advancing field, providing an in-depth examination of the fundamental principles, technical challenges, and future prospects of nanofluids in thermal and acoustic energy transfer applications.

This book is meticulously organized to assist both novices and experienced professionals in comprehending the fundamental principles of acoustic cavitation and its significance in nanofluid dynamics. The first chapters establish a solid foundation by presenting essential concepts such as acoustic cavitation, ultrasonography principles, and the dynamics of acoustic bubbles. These chapters are vital for understanding the influence of cavitation phenomena on the performance of nanofluids in heat and mass transfer processes – a recurring debated issue throughout the course.

Subsequent to those chapters addressing the fundamentals, this book progresses into the development and engineering of nanofluids, with a specific focus on ultrasonically assisted production methods. These sections elucidate the complex mechanisms via which sonic cavitation improves the thermophysical characteristics of nanofluids. These chapters provide a link between theoretical and practical insights on utilizing ultrasonication to enhance the stability, dispersion, and overall efficacy of nanofluids in practical applications. Case studies are incorporated to demonstrate realistic instances of how ultrasonically assisted approaches enhance nanofluid performance, specifically in heat transfer and improved overall oil recovery and such contexts.

This book is distinguished by its integrative approach, which harmonizes a robust theoretical base with realistic industrial applications. Chapters focusing on hydrodynamic and acoustic cavitation in microchannels, together with the augmentation of heat transmission through nanofluids, offer critical insights into the potential of nanofluids to enhance energy systems such as heat exchangers, boilers, and microfluidic devices. Furthermore, this book examines innovative applications, including the utilization of nanofluids in wastewater treatment and enhanced oil recovery, where the interaction of cavitation with nanofluid technology is expected to yield substantial improvements in process efficiency.

This book addresses the inherent issues and constraints related to nanofluids, in addition to providing technical insights. In-depth analyses of cost calculation, scalability, and life cycle assessment offer readers a thorough comprehension of the economic and environmental dimensions of nanofluid technology and its implementation. The use of a strength, weakness, opportunities, and threats (SWOT) analysis encourages researchers and industry professionals to evaluate the long-term feasibility and hazards linked to the deployment of nanofluids across diverse sectors.

Thermo-Acoustics of Nanofluids and Transfer Processes stands out as a comprehensive and state-of-the-art resource, enriched with carefully illustrated figures and tables that enhance the reader's understanding of these complex concepts. This visual clarity is one of the most distinguishing features of this book, making it an invaluable tool for students, researchers, scientists, academicians, and industry professionals seeking to understand and to exploit the vast potential of nanofluids and their wide-ranging applications. Published by CRC Press, this book is particularly valuable for those in the fields of chemical engineering, nanotechnology, environmental science, and renewable energy.

My compliments to Prof. Shriram Sonawane and his team on this very much-needed review.

Prof. A. B. Pandit
Vice-Chancellor,
Institute of Chemical Technology, Matunga, Mumbai, India

Preface

Nanofluids are gaining increasing recognition among scientists and researchers worldwide. This novel heat transfer medium improves the thermal conductivity of any working fluid by suspending small amounts of nanoparticles in the base fluid, leading to significant improvements in heat transfer in a variety of real-time industrial applications. *Thermo-Acoustical Studies of Nanofluids* delves into the fundamental physical principles and theoretical aspects of acoustics of nanofluids, with a focus on the mathematical rigor, behind their applications in fields such as heat transfer, petroleum science and technology, wastewater treatment, food processing, and hydrogen production. This work brings together expert contributions from around the world to provide a comprehensive understanding of the systems.

This book provides a rigorous framework to aid readers in building innovative nanofluid-based devices, covering essential topics such as the study and measurement of thermophysical characteristics, convection, and heat transfer equipment performance. There are various books available on the market on the heat transfer application of nanofluids, but none incorporate the effect of acoustic cavitation in the industrial application of nanofluids.

Designed to be accessible, this book has no prerequisites, making it suitable for scientists and scholars from various disciplines of research and technology. It presents foundational knowledge in engineering acoustics and introduces nanofluids and their wide-ranging applications, enabling readers from any background to explore how nanofluids can be employed in innovative ways.

The following key features and contents of this book are designed to provide readers with valuable insights and practical knowledge:

i. Fundamentals of acoustic cavitation and nanofluids
ii. Application of ultrasound-assisted nanofluids in heat transfer, enhanced oil recovery, and wastewater treatment
iii. Process analysis and optimization
   - Advanced numerical approach
   - Life cycle analysis
   - Industrial scale-up challenges
   - Future works

In addition to these core topics, this book provides a thorough exploration of the effects of acoustic cavitation on the performance of nanofluids, shedding light on how cavitation enhances heat and mass transfer efficiency in industrial processes. It delves into the synergies between ultrasound-assisted techniques and nanofluids, offering practical insights into how these combined technologies are driving innovations in fields such as enhanced oil recovery, wastewater treatment, and thermal management.

Each chapter is supported by comprehensive theoretical models and validated through experimental data, ensuring that the reader gains a balanced understanding of both the foundational concepts and their practical applications. The inclusion of advanced numerical methods provides valuable tools for process simulation and optimization, while the discussions on industrial scale-up and life cycle analysis ensure relevance to real-world challenges.

Moreover, future research directions are highlighted throughout the book, offering readers a glimpse into emerging trends and potential breakthroughs in the application of nanofluids and ultrasound technologies in industrial processes. This book will be invaluable for researchers and professionals in both industry and academia who are involved in the design and application of nanofluids in industrial processes. It is particularly aimed at:

1. Undergraduate, postgraduate, and doctoral students in the field of nanotechnology/chemical engineering and aspiring researchers within a broad domain of nanotechnology/chemical engineering.
2. Faculty and staff from reputed academic and technical institutions.
3. Executive engineers and researchers in the chemical and biomedical industries.
4. Government organizations, including R&D laboratories focused on nanotechnology and chemical processes.
5. Professionals working in chemical, materials, and mechanical engineering, nanoscience, soft matter physics, and chemistry.

About the Authors

Shriram S. Sonawane is a professor in the Department of Chemical Engineering at the Visvesvaraya National Institute of Technology in Nagpur, India and have more than 20 years of teaching and research experience. His research areas include polymer nanocomposites, nanofluids, nano-separation technology, and process modeling and simulation. Based on his expertise, he has edited three books: *Novel Approaches to Wastewater Treatment and Resource Recovery, Applications of Nanofluids in Chemical and Bio-medical Process Industry,* and *Nanofluid Applications for Advanced Thermal Solutions.* He also wrote three books as an author with the publisher Taylor & Francis (CRC Press).

He has received a fellow award from the prestigious Maharashtra Academy of Science and is also a fellow of IIChe, IWWA WRJ, RJCE, etc. He has more than 12 Indian patents to his credit. He has published more than 200 research papers in various national and international journals of repute, receiving more than 6000 citations for his work on. He has also presented more than 150 research papers in various nationals and international conferences and organized various national and international conferences. Many more awards to his credit in the field of science and technology, which include Best Engineers Award and Best Researcher's Award, and nominated for the prestigious "Shanti Swaroop Bhatnagar Award for two consecutive years (2020 and 2021). He has developed various high-performance nanofluids for application in pool boiling, heat exchange, solar panels, wastewater treatment, petroleum science and technology, membrane technology, and car radiator applications. With his team from the Department of Chemical Engineering, he has succeeded in working out a method to improve the production of bio-hydrogen from complex distillery wastewater. He has also developed novel packing materials for extraction operations. He is also a reviewer for journals such as the *Canadian Journal of Chemical Engineering, Institution of Engineers, Chemical Engineering and Technology, Ultrasonic Sonochemistry,* and many more.

Manjakuppam Malika is an assistant professor in the Department of Chemical Engineering, B V Raju Institute of Technology (BVRIT), Narsapur, Telangana, India. She was recognized in Stanford University's "World's Top 2%" scientists list for the year of 2024. Her current research focuses on nanofluid applications in chemical engineering and technology.

1 Introduction to Acoustic Cavitation

1.1 WHAT IS ACOUSTIC CAVITATION?

1.1.1 DEFINITIONS

- Acoustics is the study of mechanical waves in a variety of media, including gases, liquids, and solids. It involves the investigation of phenomena such as vibration, sound, ultrasound, and infrasound.
- Cavitation is the breakdown of bubbles.
- Ultrasound is the term used to describe sound waves that have frequencies greater than the maximum limit of human hearing, which is typically above 20 kHz. Ultrasound waves are utilized in many domains because of their capacity to travel through different substances and engage with materials in distinctive manners.
- Acoustic cavitation is the process in which bubbles form, expand, and then collapse violently in a liquid when exposed to sound waves. This phenomenon occurs during the application of high-intensity ultrasound, where the low-pressure portion of the sound wave creates empty spaces or bubbles in the liquid.

An acoustic wave, commonly referred to as *sound*, is formed when pressure oscillations propagate through a medium, such as a liquid, gas, or solid, at the speed of sound. Ultrasound is defined as a sound that cannot be heard and has a frequency of pressure oscillation exceeding 20 kHz (20,000 cycles/s). Ultrasonography is commonly defined as an acoustic wave with a frequency over 10 kHz for simplicity (Figure 1.1). It is noteworthy that certain adolescents possess the ability to perceive ultrasonic waves with a frequency of roughly 20 kHz. In ultrasonic experiments, it is common to transmit audible sound at a frequency below 20 kHz into the surrounding environment. This phenomenon arises due to certain nonlinear events. To safeguard one's health, it is advisable to use earplugs or headphones during tests that employ intense ultrasound at relatively low frequencies.

A waveform (Figure 1.2) is a visual depiction of a signal or sound as it propagates through a medium over a period of time. Waveforms can be differentiated from one another based on the following features.

- Amplitude
- Frequency
- Wavelength

DOI: 10.1201/9781003594949-1

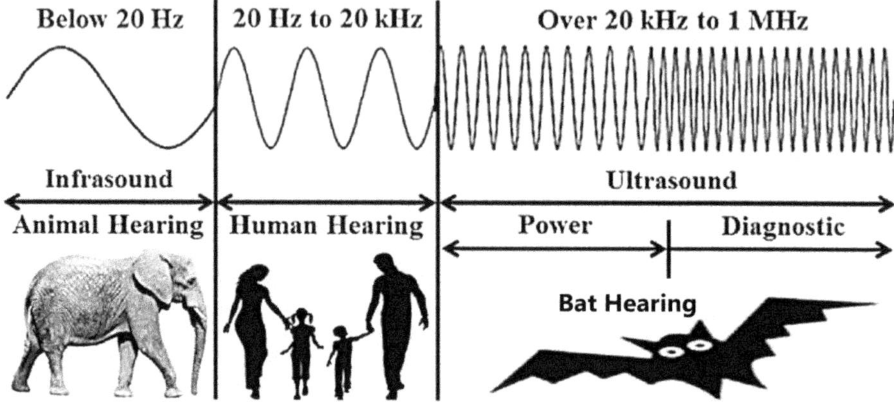

FIGURE 1.1   Sound wave frequency range.[1]

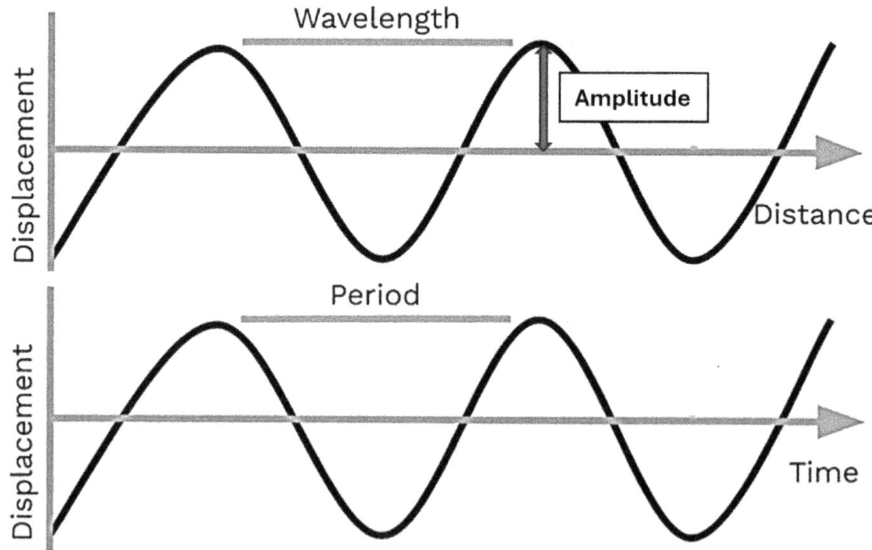

FIGURE 1.2   Characteristics of sound wave in terms of displacement with respect to distance and time.

- Acoustic period
- Acoustic pressure amplitude

1.1.2 AMPLITUDE (A)

Definition: Amplitude refers to the vertical distance measured from the centerline or x-axis. The signal intensity increases as the distance from the centerline increases.

Units: Amplitude is commonly quantified in Pascals (Pa) when referring to sound waves in a physical medium.

Impact: Waves with a greater amplitude contain a greater amount of energy and are interpreted as sounds that are louder. In contrast, waves of lesser amplitude are less intense.

Mathematically, if the displacement of the wave is given by:

$$y(t) = A \sin(\omega t + \phi) \tag{1.1}$$

where y is the displacement of wave, ω is the angular frequency, t is the time, and ϕ is the phase angle.

1.1.3 FREQUENCY (*f*)

Definition: *Frequency* (*f*) refers to the speed at which a signal repeats a complete cycle of positive and negative amplitude, covering 360°, within a specific time period.

Units: Sound waves are quantified by their frequency, which is measured in Hertz (Hz) and determined based on a period of 1 s.

Impact: A greater number of cycles or repetitive vibrations occurring within a specific time span indicate a higher frequency.

Mathematically, frequency is calculated by,

$$\text{Frequency } (f) = \frac{\omega}{2\pi} \tag{1.2}$$

1.1.4 WAVELENGTH (*λ*)

Definition: *Wavelength* (*λ*) is the quantification of the extent of displacement relative to distance. Wavelength refers to the precise measurement of the physical distance from the beginning to the end of a single wave cycle. This measurement can be determined by employing mathematical methods and calculated by using Equation (1.3).

$$\text{Wavelength } (\lambda) = \frac{\text{Speed of light}}{\text{Frequency}} = \frac{c}{f} \tag{1.3}$$

Measurement units: Expressed in meters (m).

Impact: The wavelength and frequency have an inverse relationship. Waves with higher frequencies possess shorter wavelengths, while waves with lower frequencies possess longer wavelengths.

1.1.5 ACOUSTIC PERIOD (*T_a*)

Definition: Acoustic period (T_a) is a quantitative measure of the displacement that occurs during a specific duration of time.

Units: Measured in seconds (s).

Impact: The period of a sound wave influences its time structure. A shorter period corresponds to a higher frequency (high pitch), while a longer period corresponds to a lower frequency (low pitch).

1.1.6 ACOUSTIC PRESSURE AMPLITUDE

Definition: Acoustic pressure amplitude is the highest level of pressure fluctuation in the medium caused by the sound wave, compared to the surrounding pressure. The term "peak deviation" refers to the maximum difference between the pressure in the medium and its normal atmospheric pressure caused by a sound wave.

Units: Expressed in Pascals (Pa).

Impact: The impact is directly correlated with the energy and intensity of the sound wave. Increased acoustic pressure amplitude leads to louder sounds and potentially more pronounced physical effects, such as cavitation.

The pressure fluctuation of a sinusoidal sound wave, denoted as $p(t)$, can be expressed as:

$$p(t) = p_a \sin(\omega t + \phi) \tag{1.4}$$

Here,

p_a = Pressure Amplitude
ω = Frequency (Hz)
ϕ = Initial Phase (Radians)

1.1.6.1 Important Points

If the frequency (f) of an acoustic wave is the number of pressure oscillations that occur in a given unit of time (s), it can be represented by Equation (1.5).

$$\text{Frequency } (f) = \frac{1}{T_a} \tag{1.5}$$

The sound velocity, or sound speed (c), can be defined as the distance traveled by a pressure disturbance in a given unit of time. The speed of sound can be calculated from Equation (1.6).

$$\text{Speed of sound } (c) = \text{Frequency} \times \text{Wavelength} = f\lambda \tag{1.6}$$

Generally, the speed of sound in dry air at room temperature is around 340 m/s, while in liquid water, it is approximately 1,500 m/s. The speed of sound in liquid water rises as the temperature increases, peaking at around 1,555 m/s at around 74°C.

1.2 HISTORICAL OVERVIEW

Acoustic cavitation and ultrasonication are closely connected phenomena that have had a profound influence on numerous scientific and industrial domains. The historical evolution of these notions showcases a captivating voyage of exploration, experimentation, and technological progress. This essay delves into the historical background of acoustic cavitation and ultrasonication, emphasizing significant milestones and contributions that have influenced our comprehension and utilization of these phenomena.

1.2.1 INITIAL OBSERVATIONS AND THEORETICAL FOUNDATIONS

1.2.1.1 Unveiling the Phenomenon of Cavitation

The occurrence of cavitation was initially recognized in the realm of hydraulic technology. During the latter part of the 19th century, engineers observed a significant occurrence of erosion and damage in ship propellers and turbines. In 1894, British engineer John Isaac Thornycroft and his colleague Sydney W. Barnaby discovered that the reason behind the problem was the creation and subsequent bursting of vapor bubbles in the water. They referred to this phenomenon as "cavitation."

1.2.1.2 The Contribution of Lord Rayleigh

In 1917, Lord Rayleigh (John William Strutt) made a notable theoretical contribution by examining the kinetics of a solitary gas bubble within a liquid. The paper he authored, titled "On the Pressure Developed in a Liquid During the Collapse of a Spherical Cavity," presented a mathematical framework that elucidated the phenomenon of cavitation. Rayleigh's equation elucidated the dynamics of bubbles and the profound pressures generated during their implosion, establishing the foundation for subsequent investigations.

1.2.2 Advancement of Ultrasonication

1.2.2.1 Initial Investigations on Ultrasonic Waves

Ultrasonication, which involves the application of high-frequency sound waves to bring about physical and chemical alterations in materials, originated in the early 20th century. In 1880, the French physicist Pierre Curie and his brother Jacques made the discovery of piezoelectricity, which refers to the ability of certain materials to produce an electric charge when subjected to mechanical stress. This discovery was essential for the advancement of ultrasonic transducers.

1.2.2.2 Langevin's Innovations

French physicist Paul Langevin, in collaboration with Russian scientist Constantin Chilowski, pioneered the development of the first functional ultrasonic transducer by utilizing quartz crystals during World War I. The Langevin oscillator, originally designed for submarine detection using sonar technology, was created. Langevin's contributions initiated the utilization of ultrasonics in many areas outside its military applications.

Progress in the study of acoustic cavitation Advancements by Milton S. Plesset and Additional Theoretical Developments. American physicist Milton S. Plesset made substantial progress in the theoretical comprehension of cavitation throughout the mid-20th century. Plesset enhanced Rayleigh's models by including aspects such as temperature effects and mass transport. His research yielded a profound understanding of the intricate dynamics of cavitation bubbles and their interactions with acoustic fields.

1.2.3 Photography that Captures Images at a Rapid Rate

Scientists were able to visually capture the dynamics of cavitation bubbles with the introduction of high-speed photography in the 1950s and 1960s. A. T. Ellis and R. T. Knapp employed high-speed cameras to directly watch the process of bubble formation, expansion, and collapse in real time. These visualizations yielded useful empirical data for validating theoretical models and gaining insight into the violent nature of cavitation.

1.3 ULTRASOUND PRINCIPLES

In simple terms, sonic cavitation is the process of creating, expanding, and collapsing small vapor-filled spaces, called bubbles, within a liquid medium under strong sonic forces. These bubbles form as a result of acoustic pressure fluctuations caused by ultrasonic vibrations traveling through the liquid. When sound waves create a decrease in pressure, the liquid undergoes tension, leading to its stretching and the formation of bubbles if the tension surpasses the cohesive forces of the liquid.

During the ultrasonication process, sound waves propagate through the liquid, creating alternating cycles of high pressure (compression) and low pressure (rarefaction). The pace of these cycles varies depending on the frequency. During the low-pressure cycle, the application of high-intensity ultrasonic waves induces the creation of minuscule vacuum bubbles or empty spaces in the liquid being subjected to sonication. These empty spaces or cavities are known as vacuum bubbles. After a high-pressure cycle, these bubbles reach their maximum volume. At this point, they can no longer absorb energy and they break apart violently as shown in Figure 1.3. Cavitation, the breakdown of bubbles in the sonicated liquid, causes significant effects, resulting in pressures reaching 1,000 atm and temperatures reaching 5,000 K. Acoustic cavitation exerts sufficient force to disintegrate small particle clusters and disperse them more uniformly throughout liquids. As a result of this spread, there is a substantial level of turbulence in the immediate surroundings. These significant "micro" events can cause polymer chains to easily break or completely destroy the cell walls of biological tissues. Furthermore, it possesses the ability to induce the breakdown of water molecules and produce highly reactive free radicals $\left(H_2O \rightarrow H^+ + OH^- \right)$, which can interact with and alter other molecules.

The expansion of these bubbles is a dynamic phenomenon that is affected by various parameters, including the frequency and strength of the sound waves, as

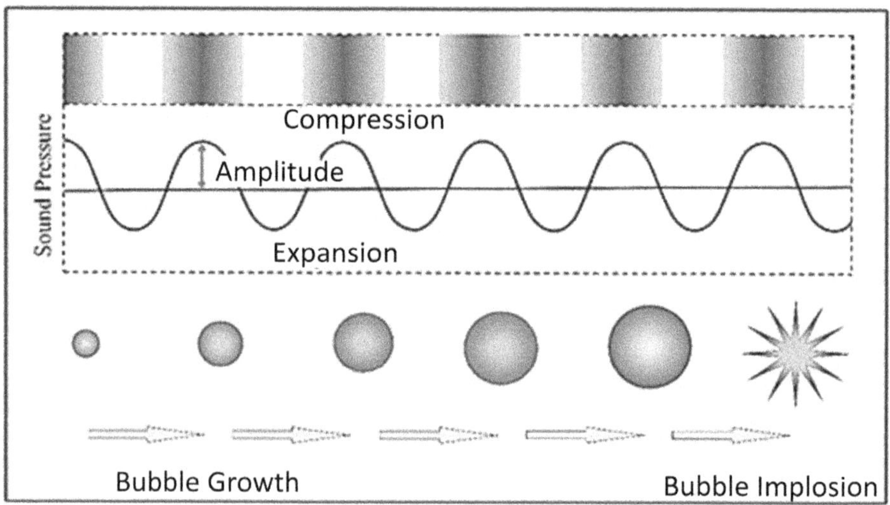

FIGURE 1.3 Ultrasonication mechanism.[2]

well as the characteristics of the liquid medium. As the bubbles increase in size, they undergo oscillations in response to the alternating pressure waves. Eventually, they reach a critical size at which they are no longer able to maintain their structure, resulting in their collapse. The phenomenon referred to as cavitation implosion results in the production of highly concentrated forces and temperatures, leading to the formation of microjets, shockwaves, and even minuscule sparks. The diversity and wide-ranging uses of acoustic cavitation are highly impressive. Within the medical field, it is useful in the areas of medical imaging and therapy. Cavitation bubbles in diagnostic ultrasound can improve picture contrast, facilitating better visibility of tissues and organs. Furthermore, in the field of therapeutic ultrasonography, the deliberate collapse of bubbles can be employed to mechanically shatter malignant cells or transport medications to precise locations within the body, providing a non-invasive alternative to conventional surgical interventions.

In addition to its applications in health, sonic cavitation is crucial in other industrial processes. In the field of sonochemistry, the intense temperatures and pressures produced when cavitation collapses aid in chemical reactions that would otherwise necessitate severe conditions, hence allowing for more effective and environmentally friendly synthesis processes. Moreover, in the context of cleaning processes, such as those using ultrasonic cleaning machines, the swift generation and collapse of bubbles generate microjets and shockwaves that remove impurities from surfaces, rendering it a very efficient technique for delicately cleaning intricate components.

Nevertheless, acoustic cavitation, despite its numerous applications, also poses obstacles and risks that require thoughtful deliberation. During the application of high-intensity focused ultrasound (HIFU) for therapeutic reasons, the release of excessive energy during the collapse of cavitation might result in undesired tissue injury or unanticipated side effects. Furthermore, in industrial operations, the erosion of equipment surfaces caused by cavitation can create operational problems and require frequent maintenance.

1.4   CAVITATION FUNDAMENTAL MECHANISMS

In the field of acoustics, bubbles are formed due to the tensile pressures exerted by regions of low pressure or low density inside the sound wave. Hydrodynamic cavitation utilizes a venturi mechanism to create bubbles by decreasing the pressure in the flow downstream. Acoustic cavitation exhibits long-lasting and steady oscillation of bubbles, in contrast to the short-lived oscillation observed in hydrodynamic cavitation or the limited number of oscillation cycles seen in laser-generated bubbles. Therefore, the oscillatory impact of sonic bubbles is more noticeable. It is noteworthy that cavitation can occur in either a homogeneous or heterogeneous manner.

Furthermore, ultrasound cavitation refers to the creation and collapse of gas bubbles in a liquid due to the use of intense ultrasound. This complex and diverse phenomenon has attracted significant attention in various scientific and technological fields. The process can be loosely classified into three distinct stages: nucleation, growth expansion, and collapse, which represent the basic mechanisms involved (Figure 1.4).

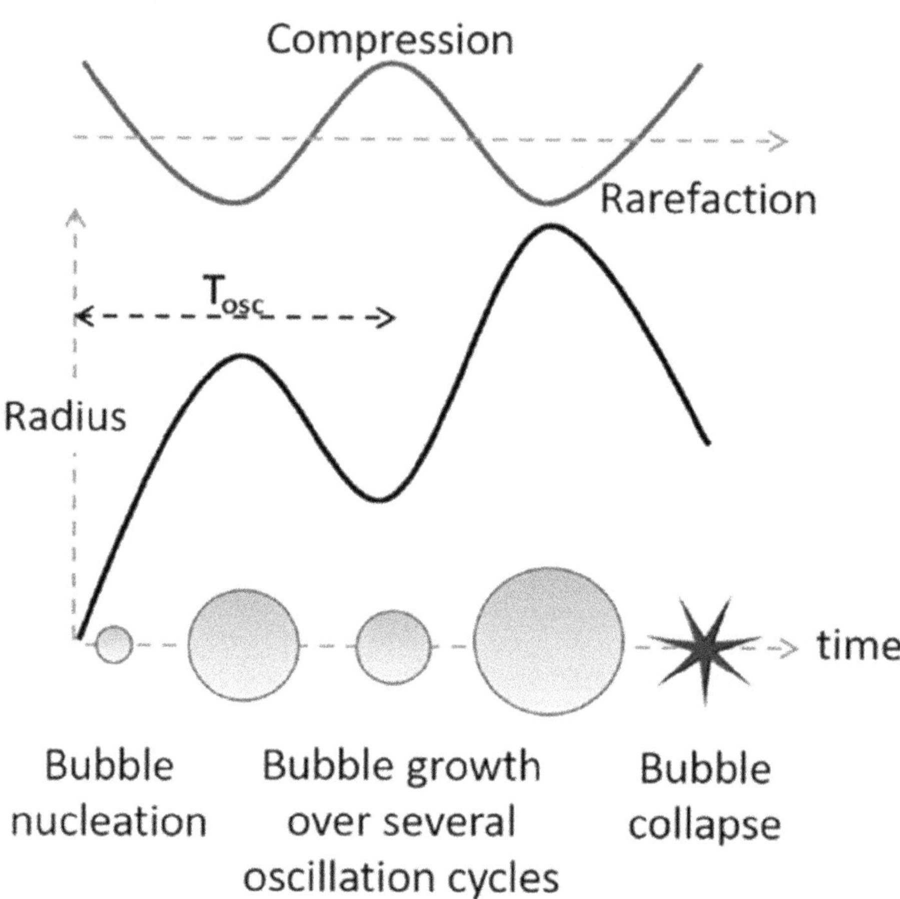

FIGURE 1.4 Fundamental cavitation principle including nucleation, bubble growth, and collapse.[3]

1.4.1 NUCLEATION

During the *nucleation* stage, the initial development of cavitation bubbles occurs. This progress can occur through several means, such as the stimulation of existing nuclei or the direct formation of new bubbles. The nuclei that act as sites for bubble formation can be small gas pockets, impurities, or irregularities within the liquid medium. These nuclei are present in the liquid media. After the ultrasonic field is applied, the sound wave's negative pressure phase creates areas of low pressure. This leads to the enlargement of the nuclei, ultimately leading to the creation of the earliest cavitation bubbles.

There are two main categories of nucleation: homogeneous and heterogeneous as shown in Figure 1.5.

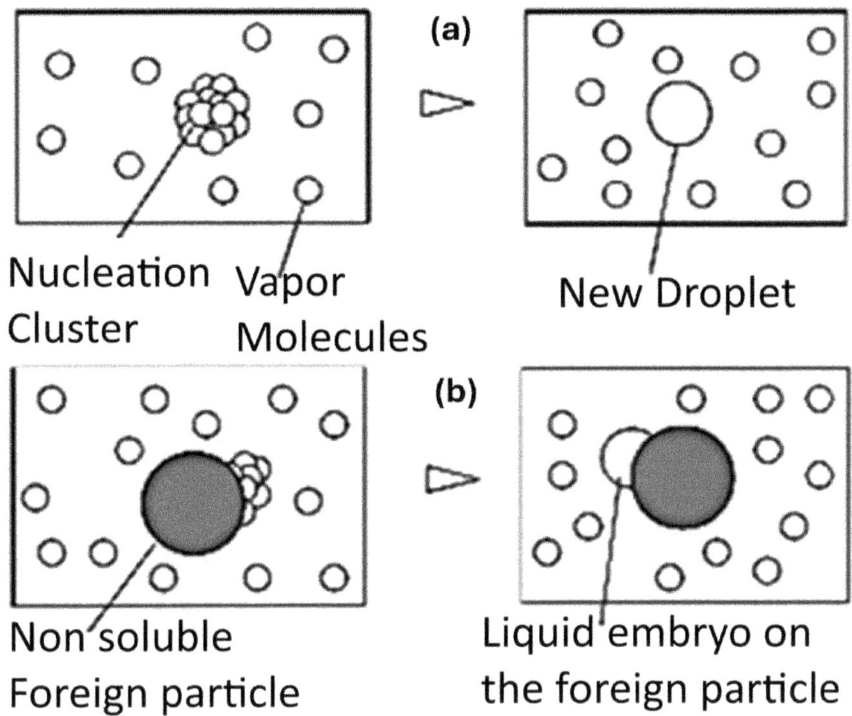

FIGURE 1.5 Illustration of (a) homogeneous and (b) heterogeneous nucleation mechanisms in ultrasound cavitation.[4]

1.4.1.1 Homogeneous Nucleation

Homogeneous nucleation refers to the process of bubble formation in a liquid that is free from any contaminants. The process necessitates a substantial expenditure of energy due to the formation of a bubble within a homogeneous liquid phase.

Mechanism: Ultrasonic waves generate alternating cycles of high pressure (compression) and low pressure (rarefaction) when they propagate through a liquid. During the rarefaction phase, the pressure in the liquid might decrease to a level below the vapor pressure of the liquid, resulting in the creation of tiny vapor cavities (nuclei).

1.4.1.2 Heterogeneous Nucleation

Heterogeneous nucleation refers to the process of nucleation occurring on a surface or interface that is different from the bulk material. Heterogeneous nucleation is more likely to occur than homogeneous nucleation due to the presence of impurities or pre-existing gas pockets that serve as nucleation sites.

Mechanism: Nucleation sites are formed by small gas pockets that become trapped on the surface of impurities or container walls. In this instance, the energy barrier for nucleation is reduced due to the ease with which the existing gas pockets can expand into cavitation bubbles as the pressure decreases during the rarefaction phase of the ultrasonic wave.

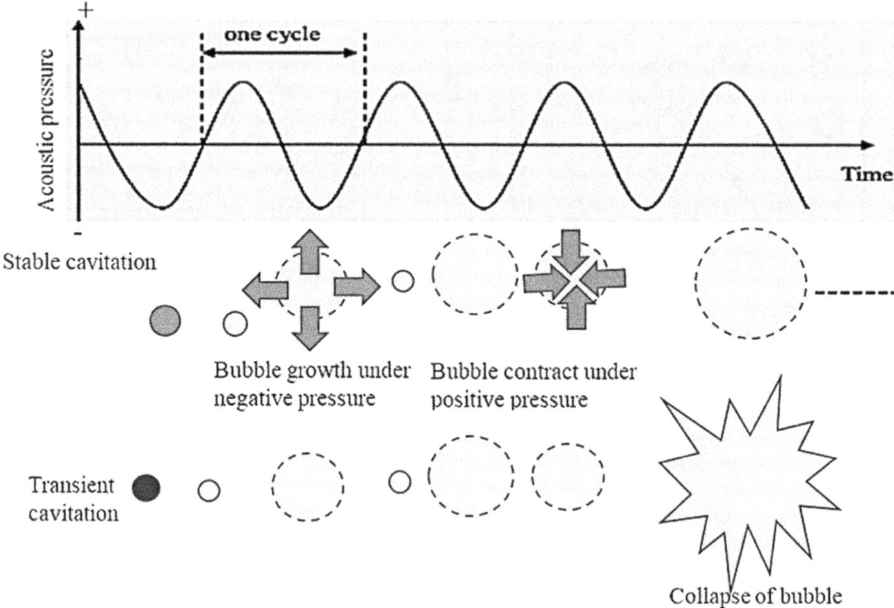

FIGURE 1.6   Illustration of bubble growth phases.[1]

1.4.2 GROWTH

After nucleation is placed, the bubbles transition into the growth phase. This phase is defined by the expansion of the bubbles during the low-pressure cycles of the ultrasonic wave as shown in Figure 1.6. The *growth* stage of cavitation bubbles involves their expansion and contraction, which is driven by the cyclic pressure fluctuations induced by the ultrasonic field. During the positive pressure phase, the bubbles are being crushed, while during the negative pressure phase, they are rapidly expanding and increasing in size. Factors such as the intensity of the ultrasonic field, the physical properties of the liquid medium, and the presence of dissolved gases can influence both the rate of bubble formation and the extent to which they grow. The growth of a bubble can be linear or nonlinear.

1.4.2.1 Linear Growth

Linear growth refers to a steady and consistent increase or progression. During the early phases, the bubbles experience linear growth as a result of the pressure difference generated by the ultrasonic wave. The rarefaction phase induces the expansion of the gas within the nucleus, leading to the development of bubbles.

Mechanism: As the pressure fluctuates, the bubbles increase in size with each cycle. The gas permeates the bubble from the adjacent liquid, augmenting the bubble's size.

1.4.2.2 Nonlinear Growth

Nonlinear growth refers to a pattern of growth that does not follow a straight or predictable path. As the size of the bubbles increases, their rate of growth becomes nonlinear. The parameters that affect it include the frequency and amplitude of the ultrasonic waves, as well as the surface tension and viscosity of the liquid.

Mechanism: In the process of nonlinear growth, the bubbles have the potential to experience fast expansion when the ultrasonic intensity reaches a certain threshold. During the rarefaction phase, the process of corrected diffusion causes gas to enter the bubble more than it leaves during the compression phase. This results in the overall development of the bubble.

1.4.3 COLLAPSE

The collapse phase is the pivotal step of cavitation, characterized by the sudden implosion of bubbles. The implosion creates localized extreme circumstances, such as elevated temperatures (around 5,000 K) and pressures (1,000 atm). The ultrasound can induce different types of cavitation including inertial and noninertial collapse. As shown in Figure 1.7.

Inertial collapse refers to the sudden loss of motion or movement in a system due to the absence or reduction of external forces or influences.

Inertial collapse, also known as transitory collapse, happens when the size of the bubble becomes larger than a specific critical radius during the low-pressure phase. The following high-pressure period induces a quick collapse of the bubble.

Mechanism: The bubble is rapidly compressed, resulting in the creation of a shock wave. During the process of collapse, there is a high level of energy concentration, which can lead to temperatures reaching several thousand Kelvin and pressures reaching several hundred atmospheres. This phenomenon is frequently utilized in applications, such as sonochemistry, to facilitate chemical reactions.

Noninertial collapse or stable collapse refers to the process of a system or structure collapsing due to forces or factors other than inertia.

Noninertial collapse, also known as stable collapse, refers to the phenomenon when a bubble undergoes oscillations in size without undergoing a catastrophic collapse. This phenomenon is commonly seen when ultrasound intensities are low or frequencies are high.

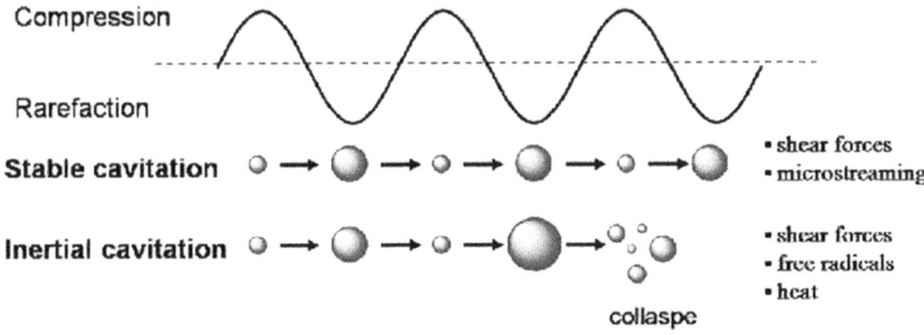

FIGURE 1.7　Illustration of bubble collapse mechanisms, highlighting inertial and noninertial collapse.[5]

Mechanism: During stable cavitation, the bubbles undergo oscillation at a size that remains relatively constant, without experiencing substantial collapse. Oscillatory motion can generate beneficial mechanical effects, such as augmenting mass transport in liquids.

1.5 ULTRASONIC TRANSDUCERS

Ultrasonic transducers are instruments that *transform electrical energy into high-frequency sound waves*, known as ultrasound, and can also convert ultrasound back into electrical energy. They play a crucial role in several applications including medical imaging, industrial cleaning, nondestructive testing, and materials processing. There are two prevalent categories of ultrasonic transducers: *direct vibratory horn transducer* and *indirect vibratory bath-type transducer*.

1.5.1 ULTRASONIC TRANSDUCERS WITH HORN-TYPE DESIGN

Horn-type ultrasonic transducers are characterized by their utilization of a horn or sonotrode to magnify and guide ultrasonic energy as shown in Figure 1.8. These transducers are employed in applications that require concentrated, powerful ultrasound.

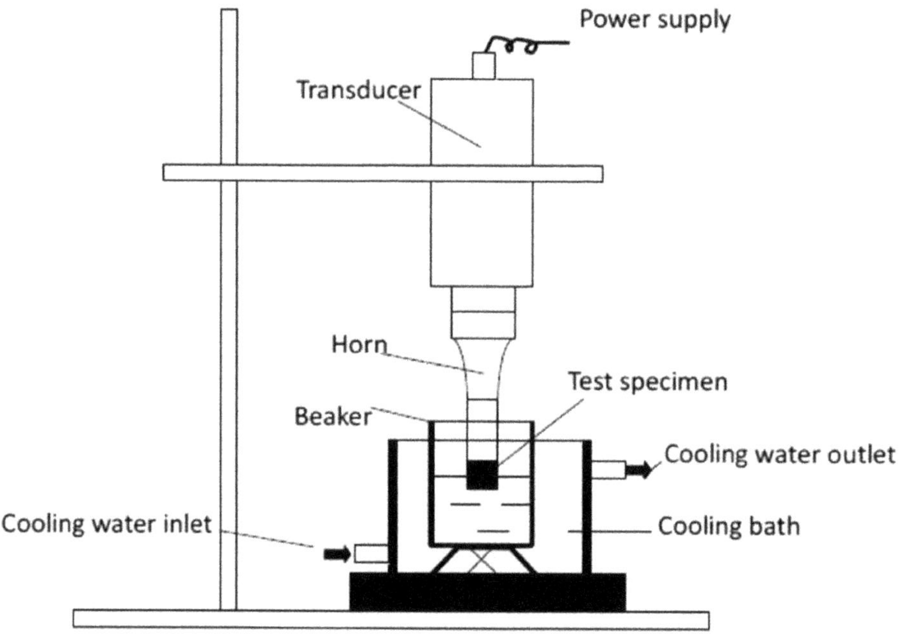

FIGURE 1.8 Diagram illustrating the structure of a horn-type ultrasonic transducer.[6] *Note*: A sonotrode is a tool that creates ultrasonic vibrations and applies this vibrational energy to a gas, liquid, solid, or tissue.

1.5.1.1 Components and Operation

1. *The transducer element*, typically made of a piezoelectric crystal or magnetostrictive material, transforms electrical signals into mechanical vibrations.
2. *Ultrasonic horn or sonotrode*, often made of titanium or aluminum, enhances and focuses the ultrasonic waves generated by the transducer element. The design of the device is intended to vibrate at the same frequency as the transducer, so increasing the magnitude of the vibrations.
3. *Booster*: Certain designs include a booster that enhances the amplification of ultrasonic waves by placing it between the transducer element and the horn.
4. *Mounting*: The transducer assembly is securely placed in a housing to guarantee stability and effective transmission of vibrations.

1.5.1.2 Applications

1. *Welding and bonding*: Horn-type transducers are commonly employed in ultrasonic welding, a process that involves the joining of materials like plastics and metals through the application of high-frequency vibrations. The horn exerts pressure and ultrasonic energy on the materials, resulting in their fusion.
2. *Cutting and slicing*: These transducers are used in ultrasonic cutting applications, where the horn oscillates to provide an accurate and neat cut through materials such as food, textiles, and composites.
3. *Medical therapeutics*: Horn-type transducers are utilized in medical settings for lithotripsy, which involves the fragmentation of kidney stones, as well as in ultrasonic surgical equipment.
4. *Nondestructive testing*: Transducers are utilized in ultrasonic testing equipment to examine materials for faults without inducing any harm.

1.5.1.3 Advantages

- *Exceptional accuracy*: The concentrated energy enables exact and meticulous application, rendering them perfect for intricate tasks.
- *High intensity*: The horn enhances the ultrasonic energy, rendering these transducers appropriate for challenging applications such as welding.
- *Versatility*: They have the ability to be utilized across a wide range of industries, encompassing fields such as medical and manufacturing.

1.5.2 Bath-type Ultrasonic Transducers

Ultrasonic transducers of the bath type are commonly employed in ultrasonic cleaning tasks. Ultrasonic waves are produced and transmitted via a liquid medium, typically water, in a tank. These waves create cavitation bubbles that effectively clean things submerged in the liquid as shown in Figure 1.9.

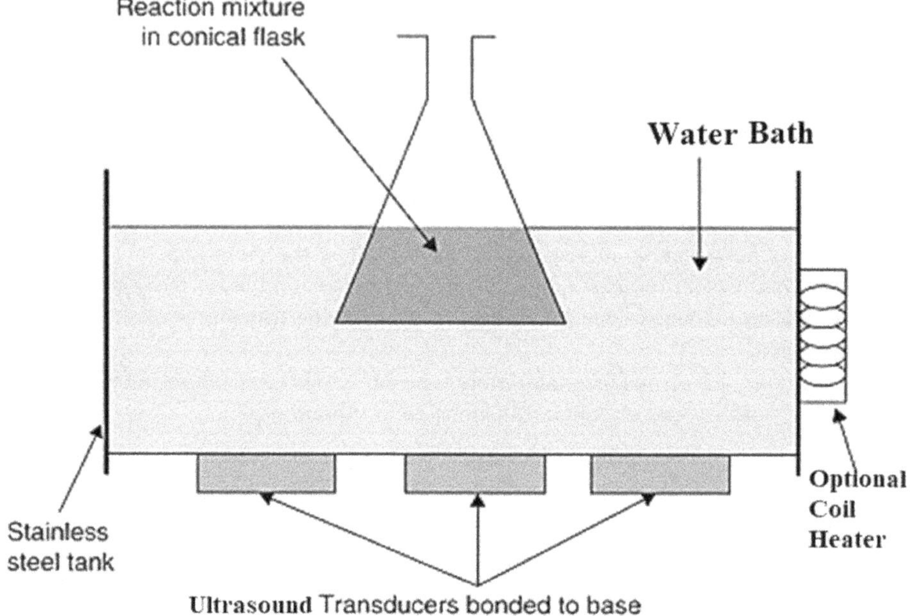

FIGURE 1.9 Illustration of a bath-type ultrasonicator.[7]

1.5.2.1 Components and Operation

1. *Transducer elements*: Typically, piezoelectric crystals are positioned on the lower or lateral surfaces of the cleaning tank to transform electrical energy into ultrasonic waves.
2. *The tank*: Typically constructed from stainless steel, serves the purpose of containing the cleaning solution and is specifically engineered to endure the powerful cavitation forces that occur during the cleaning process.
3. *Generator*: The generator provides the electrical signal that propels the transducer elements. The device has the capability to modify both the frequency and power of the ultrasonic waves.
4. *Heater*: Numerous bath-type systems use a heating mechanism to sustain the cleaning solution at an ideal temperature, hence improving the effectiveness of the cleaning process.

1.5.2.2 Applications

1. *Industrial cleaning*: Bath-type ultrasonic transducers are employed for the purpose of cleaning parts and components in several industries, including automotive, aerospace, and electronics. They possess a high level of efficacy in eliminating pollutants such as oils, greases, and particles.
2. *Sterilization of medical instruments*: These transducers are extensively employed in hospitals and laboratories to sterilize surgical instruments and laboratory equipment, guaranteeing sterility and safety.

3. Jewelry and watch cleaning services are highly sought after in the jewelry industry, particularly for the meticulous cleaning of delicate pieces that are challenging to clean manually.
4. *Laboratory application*: Bath-type ultrasonic cleaners are utilized in research environments for the purpose of removing dissolved gases from liquids, cleansing glassware, and preparing samples.

1.5.2.3 Advantages

- *Uniform cleaning*: The ultrasonic waves travel through the liquid, guaranteeing consistent and comprehensive cleaning of all surfaces of the submerged objects.
- *Delicate item-friendly*: The cavitation process is both efficient and gentle, rendering it appropriate for cleaning delicate things such as electrical components and jewelry.
- *Batch cleaning*: This process allows for the simultaneous cleaning of multiple items in a single operation, resulting in improved efficiency.

1.5.2.4 Comparative Analysis

Frequency and Power:

- *Horn type*: Generally, functions at higher frequencies (20–70 kHz) and power levels, making them well-suited for jobs that require concentrated and powerful energy.[8]
- *Bath type*: Functions within a frequency range typically between 20 and 40 kHz, specifically designed to generate cavitation in liquids with the intention of facilitating cleaning processes.

Design and Construction:

- *Horn type*: Encompasses a sophisticated design consisting of a transducer element, booster, and horn. The design prioritizes the optimization of both the magnitude and accuracy of the ultrasonic waves.
- *Bath type*: This design aims to simplify the integration of the transducer elements into a tank filled with liquid, ensuring that ultrasonic energy is evenly distributed.

Applicability Assessment:

- *Horn type*: Ideal for applications that necessitate focused and powerful ultrasonic energy, such as welding, cutting, and precision surgical treatments.
- *Bath type*: Suitable for cleaning tasks that necessitate immersion in a liquid bath, ensuring consistent cleaning throughout the whole surface of the objects.

1.6 APPLICATIONS OF ACOUSTIC CAVITATION

Acquiring knowledge and mastery over sonic cavitation has resulted in the development of numerous practical applications in diverse fields:

Sonochemistry: The use of high temperatures and pressures induced by cavitation might facilitate chemical processes that would otherwise necessitate extreme conditions. These applications encompass the synthesis of nanoparticles, the delivery of drugs, and the treatment of trash.

Medical ultrasonics: In the field of medicine, cavitation is utilized for therapeutic applications in medical ultrasonics. HIFU employs cavitation to noninvasively eradicate malignant tissues. In addition, microbubble contrast agents enhance the effectiveness of ultrasonic imaging.

Cleaning and material processing: Ultrasonic cleaning devices utilize cavitation to eliminate impurities from surfaces during the process of cleaning and material processing. Bubbles collapsing inwardly create microjets and shock waves that remove dirt and debris. Furthermore, cavitation is employed in procedures such as ultrasonic drilling and welding.

Marine engineering: Comprehending cavitation is essential for the construction of marine propellers and hydraulic turbines in the field of marine engineering. It is crucial to mitigate the effects of cavitation in order to ensure the longevity and efficiency of these components, as it can cause substantial damage to them.[9]

1.7 CHALLENGES AND FUTURE DIRECTIONS

Although there has been some advancement, there are still multiple obstacles that persist in the field of acoustic cavitation research. Forecasting and managing the occurrence of cavitation in intricate systems, such as turbulent flows, continues to be challenging. Moreover, the exact mechanisms that explain phenomena such as sonoluminescence are not yet comprehensively known.[10]

Future research endeavors to further our comprehension of cavitation at the molecular scale, potentially utilizing advancements in computational fluid dynamics and molecular dynamics simulations. Moreover, the integration of fluid mechanics, materials science, and biomedicine through interdisciplinary approaches is expected to generate novel understandings and practical uses.[11]

1.8 SUMMARY

- *Acoustic cavitation*: It is the process where bubbles form, expand, and then collapse violently in a liquid as a result of the forces produced by ultrasonic vibrations.
- *Mechanism*: Ultrasonic waves induce alternating cycles of high and low pressure when propagated through a liquid. During the low-pressure phase, the formation of tiny bubbles filled with vapor can occur, which subsequently expand and collapse forcefully during the high-pressure phase.
- There are two types of cavitation: steady cavitation, where bubbles oscillate over multiple cycles before collapsing, and transient cavitation, when bubbles rapidly expand and collapse during a single cycle.
- *Principles*: Ultrasound utilizes sound waves that have frequencies higher than what can be heard by humans, specifically above 20 kHz. These waves propagate in a medium by inducing vibrations in the particles.

- The piezoelectric effect is commonly employed in ultrasonic transducers, where materials such as quartz or ceramics produce ultrasonic waves in response to an alternating electrical field.
- Key ultrasonic equipment consist of transducers, which convert electrical energy into sound waves; generators, which supply electrical energy; and amplifiers, which enhance the signal's power. Ultrasound equipment in medical applications incorporate image display systems.

REFERENCES

1. Al-Hilphy, A. R., Al-Temimi, A. B., Al Rubaiy, H. H. M., Anand, U., Delgado-Pando, G., & Lakhssassi, N. (2020). Ultrasound applications in poultry meat processing: A systematic review. *Journal of Food Science, 85*(5), 1386–1396.
2. Shojaeiarani, J., Bajwa, D., & Holt, G. (2020). Sonication amplitude and processing time influence the cellulose nanocrystals morphology and dispersion. *Nanocomposites, 6*(1), 41–46. https://doi.org/10.1080/20550324.2019.1710974.
3. Mondal, J., Lakkaraju, R., Ghosh, P., & Ashokkumar, M. (2021). Acoustic cavitation-induced shear: A mini-review. *Biophysical Reviews, 13*(6), 1229–1243. https://doi.org/10.1007/s12551-021-00896-5.
4. Sakrani, S., Jie, L. Q., & Wahab, Y. (2005). The formation of nanoscale clusters–nanofilms/quantum dots predicted using a capillary model of nucleation. *Malaysian Journal of Fundamental and Applied Sciences, 1*, 1.
5. Zhang, C., Li, Y., Ma, X. et al. (2021). Functional micro/nanobubbles for ultrasound medicine and visualizable guidance. *Science China Chemistry, 64*, 899–914. https://doi.org/10.1007/s11426-020-9945-4
6. Wan, T., Ning, X., Hanjie, S., & Xingyue, Y. (2016). The effect of chloride ions on the corroded surface layer of 00Cr22Ni5Mo3N duplex stainless steel under cavitation. *Ultrasonics Sonochemistry, 33*, 1–9. https://doi.org/10.1016/j.ultsonch.2016.04.019
7. Paniwnyk, L. (2014). Application of ultrasound. In D.-W. Sun (Eds.), *Emerging Technologies for Food Processing* (2nd edn., pp. 271–291). Amsterdam: Elsevier. https://doi.org/10.1016/B978-0-12-411479-1.00015-2
8. Malika, M., & Sonawane, S. S. (2021). Low-frequency ultrasound assisted synthesis of an aqueous aluminium hydroxide decorated graphitic carbon nitride nanowires based hybrid nanofluid for the photocatalytic H_2 production from Methylene blue dye. *Sustainable Energy Technologies and Assessments, 44*, 100979.
9. Malika, M., & Sonawane, S. S. (2021). Statistical modelling for the Ultrasonic photodegradation of Rhodamine B dye using aqueous based Bi-metal doped TiO2 supported montmorillonite hybrid nanofluid via RSM. *Sustainable Energy Technologies and Assessments, 44*, 100980.
10. Malika, M., & Sonawane, S. S. (2021). Application of RSM and ANN for the prediction and optimization of thermal conductivity ratio of water based Fe2O3 coated SiC hybrid nanofluid. *International Communications in Heat and Mass Transfer, 126*, 105354.
11. Malika, M., & Sonawane, S. S. (2021). A comprehensive review on the effect of various ultrasonication parameters on the stability of nanofluid. *Journal of Indian Association for Environmental Management* (JIAEM), *41*(4), 19–25.

2 Foundations of Engineering Acoustics

2.1 FOUNDATION OF ACOUSTICS

Acoustics is a field of physics that focuses on the examination of sound, encompassing its generation, propagation, and impacts. The subject matter covers a broad spectrum of subjects, ranging from the fundamental principles of sound waves to their practical use in technology, medicine, and many scientific disciplines. Sound is an inevitable element in our world. Deep subterranean caves are devoid of light; however, potholers are able to perceive the sound of their own bodily movements. Sound is the only form of energy that can travel vast distances through water, which is why animals connect with one other in the deep, dark depths of the ocean. Atoms exist in isolation in cosmic space and extreme vacuums on Earth, which significantly decreases the likelihood of interaction and sound. Sound serves as a primary means of communication for humans, higher animals, and domesticated animals. Evolution has rendered certain noises enjoyable while classifying others as hazardous. The inexplicable emotional influence of music continues to be universally loved by humans. Our ears, in contrast to our eyes, possess the ability to detect sound from all directions, rendering them our primary means of receiving warnings, even during periods of sleep.

Sound is fundamental to our existence. Although audio engineers primarily focus on sound capture and reproduction, it is important for engineers in other industries to also have a professional interest in sound. The response is twofold. Sound can be utilized for diverse technical purposes, as demonstrated subsequently in this chapter. Engineers strive to mitigate the detrimental psychological and physiological effects of excessive sound on humans by developing quieter technologies, equipment, and systems, or implementing noise management techniques. The detrimental impacts of excessive sound on hearing, stress levels, rest and sleep quality, task performance, and verbal and musical communication are widely recognized and supported by recorded evidence.

Noise has emerged as a vital factor in the marketability and competitiveness of industrial products such as cars and washing machines, as shown in advertising. Various objects are required to adhere to legal and regulatory standards for noise emissions in businesses, families, and the environment. Failing to achieve these requirements has significant commercial consequences. Commercial use of airplanes is only allowed provided they meet rigorous environmental noise restrictions and receive certification. For a vehicle to be permitted on the road, it must adhere to the prescribed noise requirements. Even train noise is now subject to noise limitations.

DOI: 10.1201/9781003594949-2

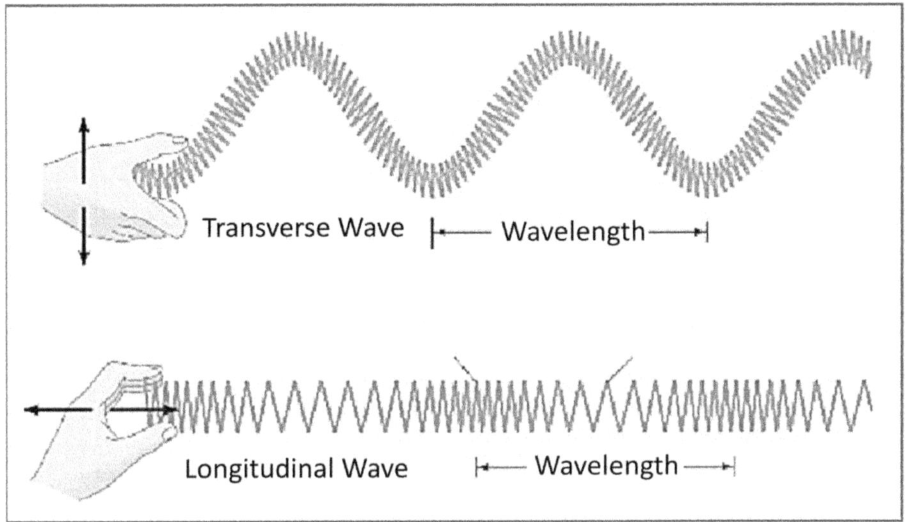

FIGURE 2.1   An illustration comparing transverse and longitudinal waves.[1]

2.2   NATURE OF SOUND

Sound is a form of mechanical wave generated from the oscillation of particles within a medium, such as air, water, or solids. These oscillations generate fluctuations in pressure that travel through the medium, resulting in our perception of sound. There are two main categories of mechanical waves: longitudinal waves and transverse waves as shown in Figure 2.1.

2.2.1   LONGITUDINAL WAVES

Longitudinal waves exhibit a particle displacement that is aligned in the same direction as the wave's propagation. These waves are comprised of alternating compressions and rarefactions of the medium they pass through.

Particle motion: The particles within the medium oscillate in a reciprocating manner, aligning with the direction of wave propagation. An example of a wave is sound waves in air, which are considered to be classic instances of longitudinal waves.

Propagation media: Longitudinal waves are capable of propagating through solids, liquids, and gases.

Particle displacement: Longitudinal waves are characterized by particle displacement occurring in the same direction as the wave's journey. This leads to areas of compression (where particles are densely packed) and rarefaction (where particles are spaced out).

Direction of propagation: Longitudinal waves exhibit a parallel movement of energy and particle motion in the same direction. This wave is characteristic of the way sound travels through different substances.

Speed of propagation: Longitudinal waves have a speed that is contingent upon the specific characteristics of the medium, including its density and elasticity. Typically, the speed of sound is greater in solids compared to liquids and greater in liquids compared to gases, owing to the density and elastic characteristics of the medium.

2.2.1.1 Important Points

- Sound waves in fluids, such as gases and liquids, are only characterized by longitudinal vibrations. Fluids are unable to enable the transmission of transverse waves due to their lack of shear strength.
- The auditory perception we experience in our daily lives is the outcome of longitudinal waves propagating through the air, distinguished by regions of high pressure (compressions) and low pressure (rarefactions).
- Sound waves in air propagate as longitudinal waves, with air molecules oscillating back and forth parallel to the direction of the wave.
- Sound in water propagates as longitudinal waves. Water's high density and flexibility enable sound to propagate at a greater speed compared to air.
- Sound waves in solid materials can propagate as both longitudinal and transverse waves, although they have distinct characteristics.
- Longitudinal waves in solids, also known as *P-waves*, consist of compressions and rarefactions that occur in the same direction as the wave's transit. These waves are commonly associated with seismic activity.

2.2.2 Transverse Waves

Transverse waves are characterized by particle movement that is perpendicular to the direction of wave transmission. These waves are composed of alternating peaks, also known as crests, and valleys, also known as troughs.

Particle motion refers to the movement of particles in a medium, which can be either up and down or side to side in relation to the direction of the wave. Example of waves: Waves on a string or electromagnetic waves, like light, are instances of transverse waves.

Propagation mediums: Transverse waves generally propagate through solids but do not propagate through fluids such as liquids and gases in the context of mechanical waves.

Particle displacement: Transverse waves are characterized by particle displacement that happens at a right angle to the direction of wave propagation. This displacement creates crests, which are the points of maximum upward displacement, and troughs, which are the points of maximum downward displacement.

Direction of propagation: The energy propagates at a right angle to the direction of particle displacement. Mechanical transverse waves predominantly manifest in solids, when particles are immobilized by robust intermolecular forces.

Speed of propagation: Transverse waves: The velocity of transverse waves is also influenced by the medium, specifically its shear modulus, which quantifies the material's stiffness. Transverse waves in solids can travel at different velocities depending on the material's stiffness.

2.2.2.1 Important Points

- Transverse sound waves in solids, known as *S-waves in seismology*, entail particle motion that is perpendicular to the direction of the wave.
- The existence of these waves is limited to solids since they necessitate the stiffness necessary to sustain motion perpendicular to the particles.

2.2.3 COMPRESSION AND RAREFACTION (EXPANSION)

Sound waves are longitudinal waves, where particle displacement occurs parallel to the direction of wave propagation. As these waves propagate across a medium (such as air, water, or solids), they generate areas of differing pressure as shown in Figure 1.2.

2.2.3.1 Compression

Compression is the term used to describe areas where particles are densely packed, leading to increased pressure. This phenomenon arises when the energy of the wave induces particles to approach each other, resulting in the formation of a more compact region within the medium.

2.2.3.2 Rarefaction or Expansion

Rarefaction is the term used to describe areas where particles are dispersed, leading to a decrease in pressure. This phenomenon happens when the energy of a wave induces the dispersion of particles, resulting in the formation of a region with lower density in the medium. Rarefaction, which is the antithesis of compression, plays an equally crucial role in the transmission of sound waves. Rarefaction is an essential factor in the propagation of sound waves, as it contributes to the periodic variations in pressure that define these waves.

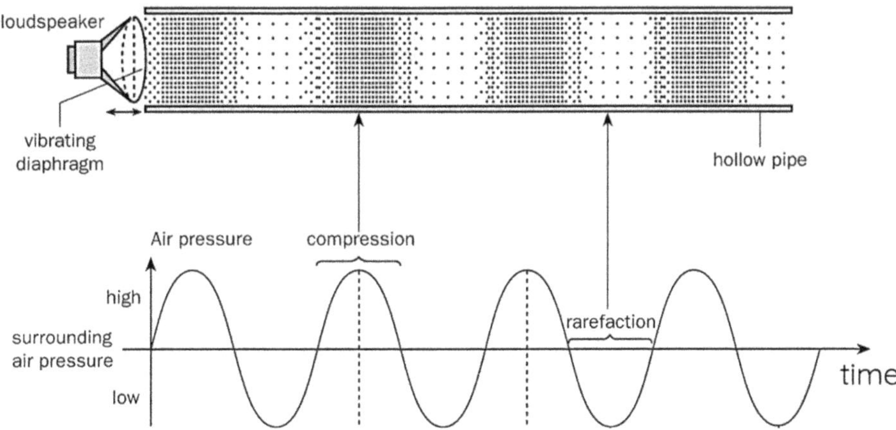

FIGURE 2.2 Representation of an acoustic wave in terms of compression and rare fraction.

2.2.3.2.1 *Mechanism of Rarefaction*

- The vibration source is responsible for the movement of a sound source, such as a vibrating string or a vocal cord. This movement causes the particles of the medium to be pushed together (compression) and pulled apart (rarefaction) in an alternating manner.
- In the process of rarefaction, particles disperse in a direction away from the source and from one another, resulting in the formation of an area with reduced pressure.
- *Wave propagation*: The less dense area moves through the substance, trailing behind the area of higher density, as the sound wave spreads.

2.2.3.2.2 *Rarefaction Characteristics*

- *Low pressure*: Rarefaction locations have a lower pressure in comparison to the surrounding areas.
- *Particle separation*: The distance between particles in the medium increases during rarefaction.
- *Energy transmission*: The sound wave's energy induces oscillation in particles, allowing the wave to propagate by alternating compressions and rarefactions.

2.3 BEHAVIOR OF SOUND WAVE

Sound waves demonstrate a range of behaviors and occurrences when they come into contact with different materials and barriers. Comprehending these phenomena is crucial for their utilization in acoustics, engineering, and research.

- *Reflection*: When a sound wave reaches a border between two distinct media, a portion of the wave is reflected back into the original medium. The angle at which a ray of light strikes a surface is equal to the angle at which it is reflected. Reflection is the fundamental concept that explains the phenomenon of echoes and is utilized in various technologies like sonar and ultrasound imaging.
- *Refraction*: The phenomenon that occurs when a sound wave transitions from one medium to another with a differing speed of sound. The alteration in velocity results in the curvature of the wave. Snell's law describes this phenomenon and is crucial for comprehending the propagation of sound in heterogeneous environments, such as the ocean or the atmosphere.
- *Diffraction*: The phenomenon of sound waves bending as they encounter obstacles or pass through apertures. The degree of diffraction is determined by the ratio of the sound's wavelength to the dimensions of the impediment or opening. This phenomenon elucidates the reason why we are able to perceive sound when it travels around corners and passes through constricted holes.
- *Absorption*: The process by which the energy of a sound wave is transformed into heat as it travels through a medium or comes into contact with

a surface. Various materials exhibit different levels of acoustic absorption, a critical factor in the construction of facilities with specialized acoustic characteristics, such as concert halls or recording studios.

- *Interference*: Arises when multiple sound waves intersect and merge. The resultant sound wave may exhibit a greater or lesser amplitude, contingent upon the phase correlation between the waves. Constructive interference leads to amplified sounds, but destructive interference can nullify the sound.
- *Doppler effect*: The Doppler effect refers to the alteration in frequency or wavelength of a sound wave as it is perceived by an observer who is in motion relative to the source of the sound. When the source and observer approach each other, the frequency of the observed sound rises, resulting in a higher pitch. When objects are moving apart from each other, the frequency of the detected waves decreases, resulting in a lower pitch. This phenomenon is frequently encountered with mobile vehicles, such as the fluctuation in frequency of a passing siren.

2.4 THE ACOUSTIC PROPERTIES OF THE BUBBLE

Bubbles, especially in liquids, play a crucial role in acoustics due to their ability to scatter and absorb sound waves. Bubbles, especially microbubbles, possess distinct acoustic characteristics that have substantial ramifications in diverse scientific and industrial contexts. The properties of the bubbles are mostly a result of the interaction between the gas inside the bubble and the liquid surrounding it. Comprehending these acoustic characteristics is essential for several applications such as medical imaging, drug administration, undersea acoustics, and others. Understanding the acoustic properties of bubbles is essential for applications ranging from underwater acoustics to medical ultrasound imaging.

2.4.1 BUBBLE DYNAMICS AND RESONANCE

The primary acoustic property of a bubble in a liquid is its resonance frequency, which is the frequency at which the bubble naturally oscillates. This resonance is primarily determined by the bubble's size and the properties of the surrounding liquid. The Minnaert frequency, named after Marcel Minnaert, who first derived it, provides an expression for the resonance frequency of a spherical bubble in an incompressible liquid. The Minnaert formula is given by:

$$\text{Resonance frequency } (f_0) = \frac{1}{2\pi}\sqrt{\frac{3\gamma P_0}{\rho R_0{}^2}} \tag{2.1}$$

where f_0 is the resonance frequency, γ is the ratio of specific heats of the gas inside the bubble, P_0 is the ambient pressure, ρ is the density of the liquid, R_0 is the equilibrium radius of the bubble.

For a single bubble in water at standard conditions, P_0 and ρ are 100 kPa and 1,000 kg/m^3, respectively. This equation shows that the resonance frequency inversely depends on the bubble radius; larger bubbles resonate at lower frequencies, while smaller bubbles resonate at higher frequencies.

Ultrasound is generated by transducers that convert electrical pulses into acoustic energy. Transducers can be classified into two categories: those that operate on the magnetostriction effect and those that operate on the piezoelectric principle. The former refers to the change in length per unit length resulting from the application of a magnetic field to a material with ferromagnetic properties. This change leads to magnetization and the subsequent production of vibrations. Nevertheless, piezoelectric transducers are the predominant choice for ultrasonic food processing. These objects are composed of two ceramic components that alter their dimensions in reaction to an electric field. Consequently, when an oscillating field is exerted, the ceramic components exhibit a consistent and predictable vertical motion.

The frequency of ultrasonic processing is crucial as it has a direct correlation with the size of the resonance bubble (R_r). Critical size is the point at which a bubble reaches its maximum size during cavitation before collapsing. R_r can be represented by Equation (2.2):

$$\text{Resonance bubble size } (R_r) = \sqrt{\frac{3\gamma P_0}{\rho \omega^2}} \tag{2.2}$$

where R_r is the resonance bubble size, γ is the ratio of specific heats of the gas inside the bubble, P_0 is the ambient pressure, ρ is the density of the liquid, ω is the angular frequency of ultrasound.

2.4.2 Scattering and Attenuation

Generally, bubbles scatter acoustic waves due to impedance mismatches between the gas inside the bubble and the surrounding liquid. The scattered intensity depends on the size of the bubble relative to the wavelength of the incident sound. For small bubbles (Rayleigh scattering regime), the scattering intensity is proportional to the sixth power of the radius and the fourth power of the frequency. For larger bubbles, comparable to the wavelength, the scattering behavior becomes more complex and requires more sophisticated models, such as the Mie theory, to describe accurately.

Attenuation, or the loss of sound intensity, occurs due to the absorption and scattering by bubbles. The presence of bubbles in a liquid can significantly increase the attenuation of sound waves, which is particularly relevant in applications such as underwater acoustics and biomedical ultrasound. The effective attenuation coefficient, which measures how quickly sound intensity decreases with distance, depends on the concentration, size distribution, and resonance frequency of the bubbles.

2.4.3 Nonlinear Effects

Bubbles exhibit nonlinear acoustic behavior, especially near their resonance frequencies. Nonlinear effects include harmonic generation, where the bubble oscillates at multiples of the driving frequency, and subharmonic generation, where oscillations occur at fractional frequencies. These nonlinearities are significant in medical

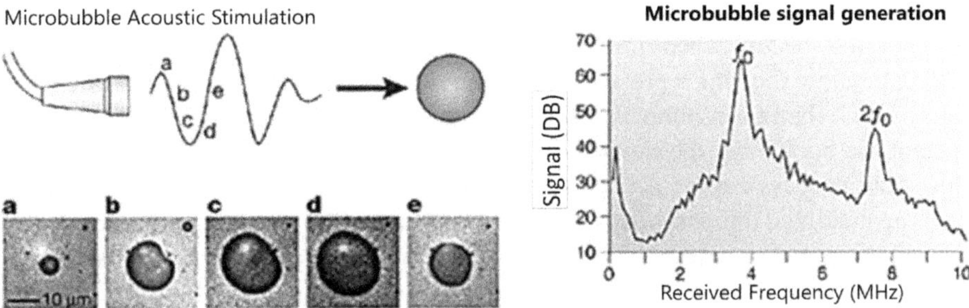

FIGURE 2.3 Acoustic simulation of microbubbles[2]: (i) the microscope images acquired at a time interval of 330 ns exhibit the oscillation in volume of a microbubble when exposed to ultrasonic waves with a frequency of 500 kHz. This oscillation occurs during both the high-pressure and low-pressure phases. The process of bubble compression and expansion takes place during distinct pressure phases of the acoustic wave, as indicated by the position of frames (a)–(e) in the schematic diagram. (ii) Microbubbles' frequency versus amplitude data shows a returning signal at both the fundamental (f_0) and second harmonic $(2f_0)$ frequencies.

ultrasound, where they enhance image contrast and enable techniques like harmonic imaging, which provides clearer images of tissues. **Figure 2.3** illustrates the non-linear behavior of microbubbles over a time interval of 330 nano seconds, showing oscillations when subjected to ultrasonic waves at a frequency of 500 kHz.

2.4.4 Applications

The acoustic properties of bubbles have numerous applications:

- *Underwater acoustics*: Bubbles influence the propagation of sound in the ocean, affecting sonar performance and underwater communication. Understanding bubble dynamics helps in predicting and mitigating these effects.
- *Medical ultrasound*: Microbubbles are used as contrast agents in ultrasound imaging. Their acoustic properties enhance the contrast between blood and surrounding tissues, improving the diagnostic capability of ultrasound.
- *Industrial processes*: Bubbles play a role in processes like cavitation, which is used in cleaning, sonochemistry, and material processing. Understanding the acoustic behavior of bubbles helps optimize these processes.

2.5 TYPES OF CAVITATION

Ultrasound utilizes sound waves at a frequency higher than the upper limit of human hearing (~20 kHz). These sound waves are generated by a ultrasound transducer, which converts electrical energy into vibrational energy. As these waves travel, they create repeated cycles of compression and rarefaction, leading to the formation of cavitation bubbles in the liquid. These bubbles can then interact with a boundary between the liquid and a solid material in various ways. *Acoustic cavitation* is a phenomenon that occurs when ultrasound waves cause the pressure to vary, sometimes dropping below the vapor pressure of the liquid. This creates small gaps filled with vapor, which then form bubbles inside the liquid. The expansion and subsequent collapse of the bubbles in the liquid medium happen very quickly, taking around 1 µs. This rapid process causes a significant increase in temperature within the bubble, reaching localized temperatures between 2,000 and 5,000 K, along with pressures of up to 1,000 atm. Consequently, the formation of hotspots leads to an increase in rates of mass and heat transfer, the advancement of sonochemical reactions, the creation of intense turbulence, and the ejection of liquid jets.

2.5.1 Summary

Cavitation is a phenomenon where vapor bubbles form in a liquid due to localized low pressures as shown in Figure 2.4. It occurs when the pressure in a liquid falls below its vapor pressure, leading to the formation of vapor-filled cavities. Cavitation can be categorized into several types, each with distinct characteristics and implications are discussed in the following sections.

2.5.2 Hydrodynamic Cavitation

Hydrodynamic cavitation (HC) occurs when the pressure in a liquid drops below its vapor pressure due to the flow dynamics. This type of cavitation is common in pumps, turbines, and propellers, where rapid changes in velocity and pressure can cause vapor bubbles to form. The cavitation-based process, as an advanced oxidation method, has emerged as a very promising technology for treating industrial effluents as shown in Figure 2.5. This is primarily owing to its cost-effectiveness in operation, improved energy efficiency, and ability to be implemented on a wide scale. HC occurs when cavities are formed due to localized pressure reductions caused by different constrictors or by the mechanical rotation of devices such as the vortex diode and other rotating-type mechanisms.

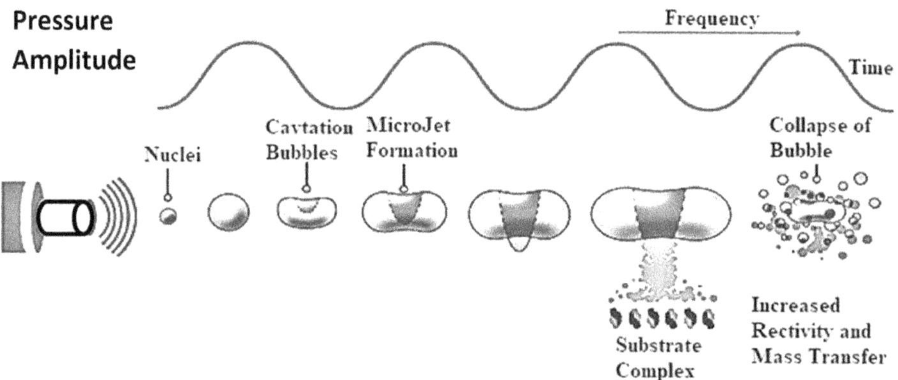

FIGURE 2.4 Schematic depiction of the ultrasonic cavitation phenomenon.[3]

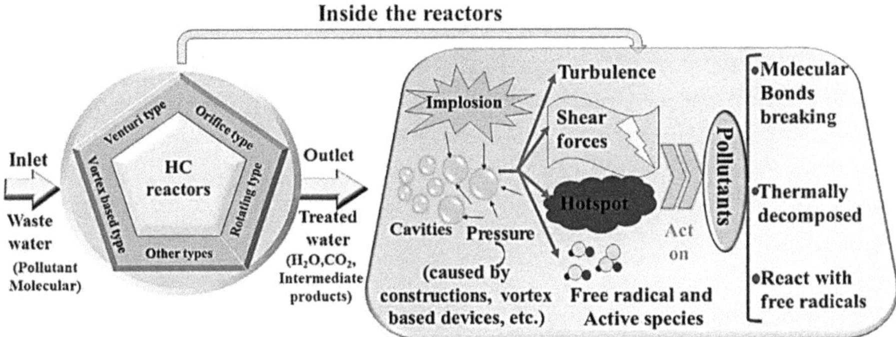

FIGURE 2.5 Different stages of hydrodynamic cavitation with different cavitational events.[4]

- *Inception and growth*: Cavitation begins when the pressure in a region of the liquid falls below the vapor pressure. Microbubbles or nuclei present in the liquid act as initiation sites for cavitation. These bubbles grow rapidly, driven by the pressure differential.
- *Collapse and effects*: When the bubbles enter regions of higher pressure, they collapse violently. The collapse generates high pressures and temperatures, producing shock waves and micro-jets that can cause significant damage to nearby surfaces. This can lead to erosion, noise, and vibration in hydraulic machinery.

2.5.3 ACOUSTIC CAVITATION

Acoustic cavitation occurs when a liquid is subjected to intense sound waves, typically in the ultrasonic frequency range. The alternating pressure waves create conditions where bubbles can form, grow, and collapse as shown in Figure 2.6. As shown in Figure 1.6, there are two types of acoustic cavitation.

- *Stable cavitation*: In stable cavitation, bubbles oscillate around an equilibrium size without collapsing. This type of cavitation is useful in applications like sonochemistry, where the oscillating bubbles facilitate chemical reactions by enhancing mass transfer and mixing.
- *Transient cavitation*: Transient or inertial cavitation involves the rapid growth and violent collapse of bubbles. This type of cavitation generates extreme conditions and is used in applications like ultrasonic cleaning, where the high-energy collapse of bubbles removes contaminants from surfaces.

2.5.4 VORTEX CAVITATION

Vortex cavitation occurs in regions of swirling flow, such as the tip of a propeller blade or in the wake of a submerged object. The low-pressure core of the vortex provides the conditions for bubble formation.

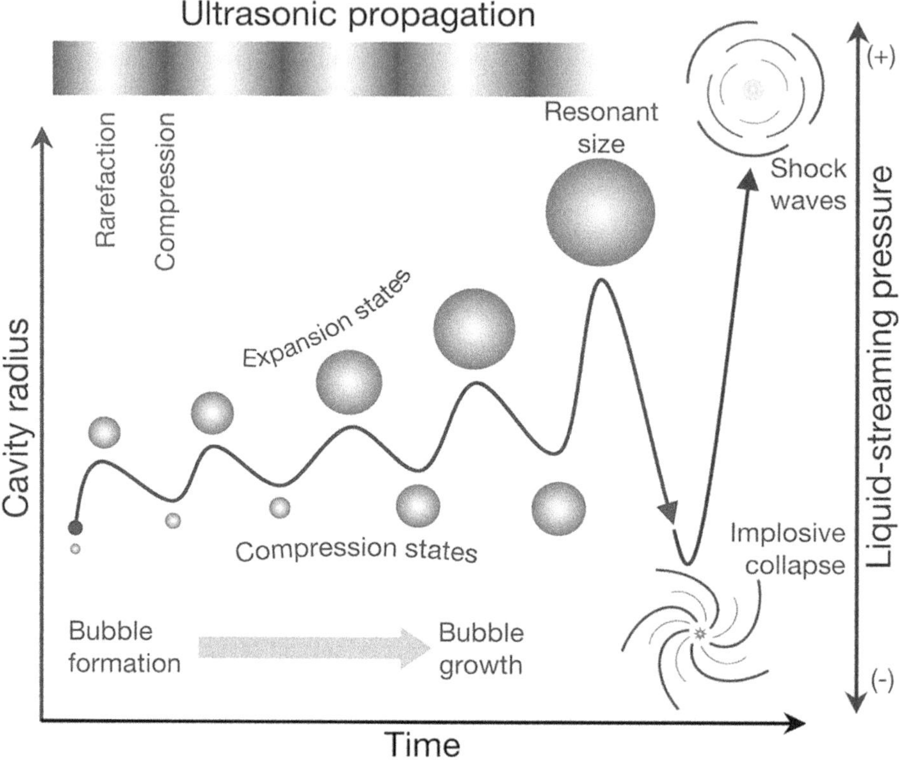

FIGURE 2.6 Schematic representation of acoustic cavitation: Bubble formation stage to implosive and collapse phase.[5]

- *Formation*: The formation of cavitation in vortices is influenced by factors like the strength and size of the vortex, the flow velocity, and the ambient pressure.
- *Impact*: Vortex cavitation can lead to the erosion of surfaces in contact with the vortices, such as propeller blades or hydraulic structures. The cyclic formation and collapse of bubbles in the vortex core can cause significant damage over time.

2.5.5 SHEET CAVITATION

Sheet cavitation, also known as cloud cavitation, occurs when vapor cavities form in a continuous sheet along a surface, such as the suction side of a hydrofoil or a propeller blade. This type of cavitation is characterized by the formation of a vapor sheet that can detach and collapse.

- *Characteristics*: Sheet cavitation starts as small cavities that coalesce into a larger vapor sheet. The sheet can oscillate, creating regions of high and low pressure.

- *Consequences*: The collapse of the vapor sheet generates pressure waves that can cause significant damage to surfaces. Sheet cavitation is associated with high levels of noise and vibration, which can affect the performance and longevity of hydraulic machinery.

2.5.6 TRAVELING BUBBLE CAVITATION

Traveling bubble cavitation involves the formation and movement of individual vapor bubbles along a surface. This type of cavitation is common in situations where bubbles nucleate on surface irregularities or impurities and then move due to the flow of the liquid.

- *Nucleation and movement*: Bubbles nucleate at points of low pressure and are carried by the flow. As they travel, they can grow and collapse, depending on the local pressure conditions.
- *Impact*: The movement and collapse of bubbles can cause localized erosion and pitting on surfaces. Traveling bubble cavitation is often observed in pumps and other fluid handling equipment, where it can lead to material degradation and reduced efficiency.

2.5.7 SUPER-CAVITATION

Super-cavitation is an extreme form of cavitation where a single large cavity envelops a submerged object, significantly reducing drag. This phenomenon is utilized in high-speed underwater vehicles and projectiles.

- *Formation*: Super-cavitation is achieved by designing objects with specific shapes and using high speeds to create a stable cavity that envelops the object.
- *Benefits and challenges*: Super-cavitation reduces frictional drag, allowing for higher speeds. However, controlling the cavity and ensuring stability is challenging. The collapse of the cavity can also generate significant forces that need to be managed.

2.6 APPLICATIONS

The understanding and control of cavitation have significant implications for various engineering fields as shown in Figure 2.7. Effective management of cavitation can improve the performance and longevity of hydraulic machinery, enhance the efficiency of industrial processes, and enable advanced technologies in underwater transportation and medical treatment.

- *Marine engineering*: Cavitation is a critical concern in marine engineering, affecting the performance and durability of propellers, turbines, and pumps. Understanding the different types of cavitation helps engineers design components that minimize cavitation damage and optimize efficiency.

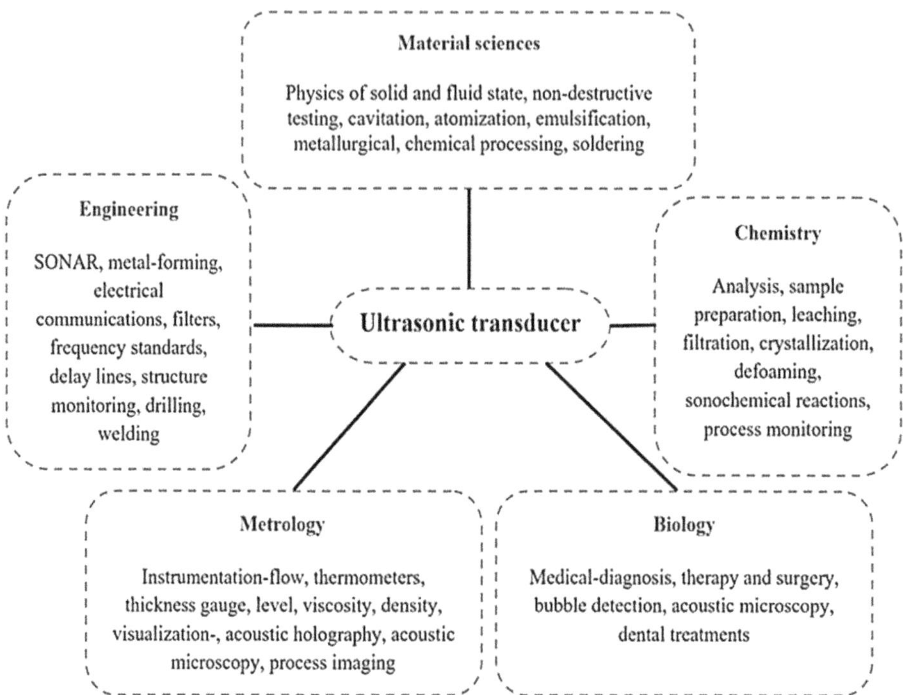

FIGURE 2.7 Applications of ultrasonication technology in various sectors.[6]

Techniques such as altering blade shapes, using cavitation-resistant materials, and implementing cavitation control measures are employed to mitigate the adverse effects.[7]

- *Medical applications*: In medical ultrasonics, controlled cavitation is used for therapeutic purposes, such as lithotripsy (breaking kidney stones) and targeted drug delivery. The ability to induce and control cavitation at specific locations enables precise treatment with minimal side effects. Understanding the dynamics of cavitation bubbles is crucial for optimizing these medical procedures.[8]
- *Industrial processes*: Cavitation plays a role in various industrial processes, such as ultrasonic cleaning, sonochemistry, and material processing. In ultrasonic cleaning, the high-energy collapse of cavitation bubbles effectively removes contaminants from surfaces. In sonochemistry, cavitation enhances chemical reactions, making processes more efficient. In material processing, cavitation is used for tasks like emulsification, homogenization, and nanoparticle synthesis.[9]
- *Environmental and energy applications*: Cavitation is explored in environmental engineering for wastewater treatment and pollutant degradation. The intense conditions generated by cavitation can break down complex pollutants, making them easier to remove or neutralize. In energy applications, cavitation is studied for its potential in processes like biofuel production and the enhancement of heat transfer in energy systems.[10]

2.7 SUMMARY

The study of engineering acoustics, particularly the acoustic properties of bubbles and the types of cavitation, reveals a complex interplay between fluid dynamics and sound. Bubbles exhibit unique acoustic behaviors, including resonance, scattering, and nonlinear effects, which have important applications in underwater acoustics, medical ultrasound, and industrial processes. Cavitation, with its various forms such as hydrodynamic, acoustic, vortex, sheet, traveling bubble, and super-cavitation, presents both challenges and opportunities across multiple fields. By understanding and controlling cavitation, engineers can improve the performance and durability of machinery, enhance medical treatments, optimize industrial processes, and develop advanced technologies. The foundational principles of engineering acoustics continue to drive innovation and solve real-world problems, demonstrating the profound impact of this interdisciplinary field.

REFERENCES

1. Zohuri, B., & Moghaddam, M. (2021). Directed energy beam weapons the dawn of a new military age. *Journal of Material Sciences & Manufacturing Research, 2*(120), 2–8.
2. Lindner, J. R. (2004). Microbubbles in medical imaging: Current applications and future directions. *Nature Reviews Drug Discovery, 3*(6), 527–533. doi:10.1038/nrd1417
3. Córdova, A., Paola, H., Helena, N., Fabián, R.-R., Cecilia, G., Carolina, A.-C., & Andrés, I. (2022). Recent advances in the application of enzyme processing assisted by ultrasound in agri-foods: A review. *Catalysts, 12*(1), 107. https://doi.org/10.3390/catal12010107
4. Wang, B., Su, H., & Zhang, B. (2021). Hydrodynamic cavitation as a promising route for wastewater treatment–A review. *Chemical Engineering Journal, 412*, 128685. https://doi.org/10.1016/j.cej.2021.128685
5. Bui, T. Q., Ngo, H. T. M., & Tran, H. T. (2018). Surface-protective assistance of ultrasound in synthesis of superparamagnetic magnetite nanoparticles and in preparation of mono-core magnetite-silica nanocomposites. *Journal of Science: Advanced Materials and Devices, 3*(3), 323–330.
6. Malek, M. N. F. A., Hussin, N. M., Embong, N. H. et al. (2023). Ultrasonication: A process intensification tool for methyl ester synthesis: A mini review. *Biomass Conversion and Biorefinery, 13*, 1457–1467. https://doi.org/10.1007/s13399-020-01100-6
7. Sonawane, S. S., & Malika, M. (2021). Review on CNT based hybrid nanofluids performance in the nano lubricant application. *Journal of Indian Association for Environmental Management (JIAEM), 41*(3), 01–16.
8. Malika, M., & Sonawane, S. S. (2021). The sono-photocatalytic performance of a novel water based Ti+ 4 coated Al (OH) 3-MWCNT's hybrid nanofluid for dye fragmentation. *International Journal of Chemical Reactor Engineering, 19*(9), 901–912.
9. Malika, M., Bhad, R., & Sonawane, S. S. (2021). ANSYS simulation study of a low volume fraction CuO–ZnO/water hybrid nanofluid in a shell and tube heat exchanger. *Journal of the Indian Chemical Society, 98*(11), 100200.
10. Sonawane, S. S., Thakur, P. P., Malika, M., & Ali, H. M. (2023). Recent advances in the applications of green synthesized nanoparticle based nanofluids for the environmental remediation. *Current Pharmaceutical Biotechnology, 24*(1), 188–198.

3 Dynamics of Acoustic Bubbles

3.1 INTRODUCTION TO DYNAMICS OF ACOUSTIC BUBBLES

Acoustic bubbles, which are used in various applications such as medical diagnostics and industrial operations, display complex behavior when exposed to acoustic fields. This study offers an in-depth examination of the underlying concepts that dictate their behavior, including the equations that govern oscillations, acoustic radiation forces, criteria for yielding, and form oscillations. The bursting of a bubble can be compared to that of a balloon, where the membrane is rapidly and completely removed. When the film of the balloon bursts, it contracts before the elastic gas inside the balloon can expand due to its elasticity. Likewise, when the bubble pops, the bubble film will quickly shrink again. As the bubble cavity returns to equilibrium, a stream of liquid is driven upwards, creating a wave of liquid.

3.1.1 FORMATION OF THE ACOUSTIC BUBBLE IN CAVITATION

Acoustic cavitation involves the application of sound waves to pre-existing cavities or bubbles in a solution, causing them to experience a sinusoidal pressure. During the negative pressure cycle, the liquid is separated at certain points in the fluid where there are gaseous pollutants. Furthermore, the process of bubble generation is recognized to take place through three distinct mechanisms.

3.1.1.1 Mechanism 1

The liquid contains bubbles that already exist and are prevented from breaking apart because of the protective layer of organic material on their surface.

3.1.1.2 Mechanism 2

The second mechanism relies on the existence of solid particles within the liquid. Nucleation begins when the liquid and gas are confined within the solids. Furthermore, it is possible that the walls of the tank or container have minuscule gaps capable of retaining gas. In a gas crevice, the pressure is lower than the pressure of the surrounding liquid. Gas diffuses into the gas pocket, resulting in the expansion of the gas pocket. Due to the effect of acoustic radiation forces, the gas pocket travels away from the fracture, leading to the creation of a bubble.

3.1.1.3 Mechanism 3

Finally, the formation of new bubbles through the disintegration of a bigger bubble.

DOI: 10.1201/9781003594949-3

3.2   DYNAMICS OF A SINGLE ACOUSTIC BUBBLE

In 1859, Besant[1] initially formulated an equation as stated below to describe the movement of the bubble wall when subjected to an acoustic field.

$$\frac{P_L - P_0}{\rho} = R\ddot{R} + 1.5\dot{R}^2 \tag{3.1}$$

where R = radius of the bubble wall as a function of time (t),

$\dot{R} = \dfrac{dR}{dt}$ = the time derivative of the bubble radius = velocity of the bubble wall,

$\ddot{R} = \dfrac{d^2 R}{dt^2}$ = second time derivative of the bubble radius (bubble wall acceleration),

P_0 : pressure in the liquid, P_L : pressure in the liquid at the bubble wall,

ρ : density of the liquid.

3.2.1   RAYLEIGH-PLESSET EQUATION

Acoustic cavitation refers to the process in which bubbles are created, enlarged, and made to oscillate within a liquid when exposed to sound waves. The Rayleigh-Plesset equation provides a basic description of bubble dynamics, including the phenomenon of violent bubble collapse. The derivation involves considering a spherical liquid volume with a radius R_L that surrounds a spherical bubble with a radius R. The center of the liquid volume is located in the center of the spherical bubble.

The behavior of acoustic bubbles, which is crucial in diverse areas like medical imaging, sonochemistry, and undersea acoustics, is determined by complex physical laws. The *Rayleigh-Plesset equation* is crucial for comprehending these dynamics. It acts as the fundamental model that explains the oscillatory behavior of gas bubbles in a liquid medium, taking into account acoustic fields, surface tension, and viscosity. Typically, a cavitation bubble contains vapor and noncondensable gases such as air. Surface tension causes the pressure within a bubble to be higher than the pressure of the liquid inside the bubble at the bubble wall. The radial oscillations of a spherical bubble in an incompressible liquid can be captured by employing the Rayleigh-Plesset equation, which is derived from the principles of fluid dynamics. Several variables, such as the gas's compressibility, the liquid's viscosity, the surface tension, the ambient pressure, and the applied acoustic pressure, are considered. The equation can be expressed as:

$$R\ddot{R} + 1.5\ R^2 = \frac{1}{\rho}\left(P_G - P_0 - P_A - \frac{2\sigma}{R} - 4\mu\frac{\dot{R}}{R} \right) \tag{3.2}$$

where R is the bubble radius, $\dot{R}$ is the velocity of the bubble wall, $\ddot{R}$ is the acceleration of the bubble wall, ρ is the density of the liquid, ρ_G is the gas pressure inside the bubble, P_0 is the ambient pressure, P_A is the acoustic pressure, σ is the surface tension of the liquid, $\sigma = 7.275 \times 10^{-2}$ (N/m) ($= $ J/m^2) for pure water at 20°C, μ is the dynamic viscosity of the liquid.

3.2.1.1 Physical Interpretation of Terms

Inertial term $(R\ddot{R})$: Describes the inertial forces caused by the acceleration of the bubble wall.

Viscous damping $\left(4\mu\dfrac{\dot{R}}{R}\right)$: The above equation accounts for the dissipation of energy caused by viscosity, where μ is the dynamic viscosity of the liquid.

Surface tension $\left(\dfrac{2\sigma}{R}\right)$: It reflects the energy needed to expand the surface area of a bubble as a result of surface tension.

Gas pressure (P_G): It refers to the force exerted by the gas molecules inside the bubble. It is determined by various parameters, including the temperature and composition of the gas.

Ambient pressure (P_0): The unchanging pressure exerted by the liquid in the immediate environment.

Acoustic pressure (P_A): The dynamic pressure applied to the bubble by the acoustic field, which changes over time and in different locations.

$$P_{\text{Acoustic}}(t) = P_0\,\sin(\omega t) \tag{3.3}$$

where ω is the angular frequency.

3.2.1.1.1 Limitations of Rayleigh-Plesset Equation

- Theoretically, it has the capability to explain significant oscillation amplitudes and the compressibility of water, as well as nonadiabatic compression and the intricate viscoelastic properties of the bubble shell.
- However, in reality, microbubbles are never seen in empty space, so further adjustments were made to the main models to consider the existence of a substrate.
- The method of images, a concept in potential flow theory, treats the substrate as an acoustic reflector and provides a convenient approach to incorporate its influence on the dynamics of the bubble.
- Additional spherical models were created and examined to study the impact of the distance between the bubble and the substrate.
- Nevertheless, the current physical issue necessitates surpassing the spherical models by taking into account the nonspherical dynamics of microbubble oscillations and release and transport of the bubbles.

3.2.2 Bubble Pressure $(P_B(T))$

The pressure within the bubble can be mathematically described by considering the gas inside the bubble to undergo adiabatic behavior as given below:

$$P_B(t) = P_G\left(\frac{R_0^{\,3}}{R^3}\right)^{\gamma} \tag{3.4}$$

where R_0 is the equilibrium radius of the bubble, γ is the polytropic exponent $= 1.4$ for adiabatic process.

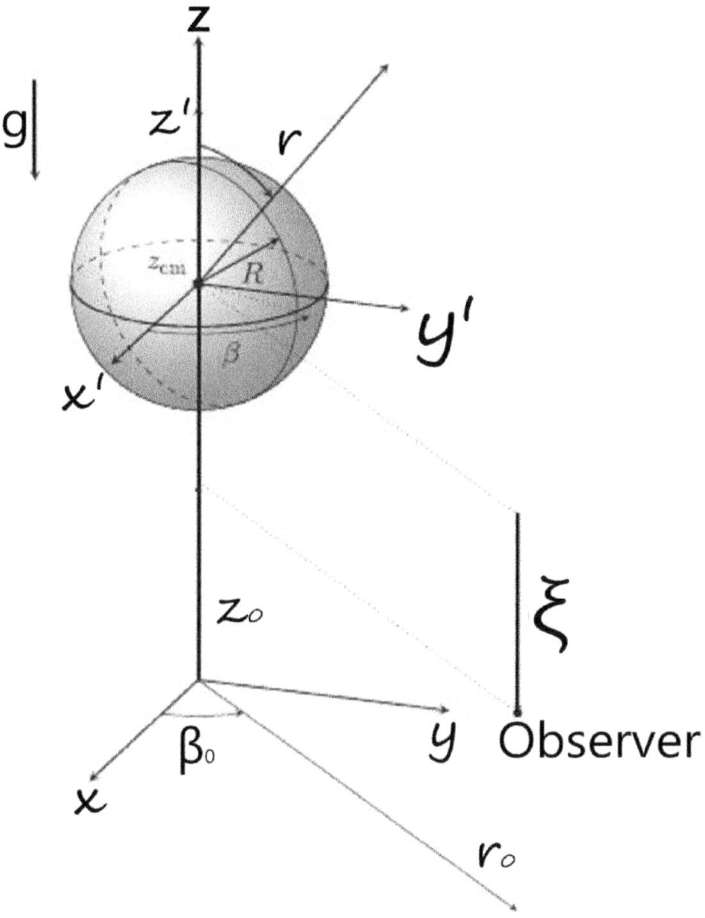

FIGURE 3.1 Schematic diagram of the acoustic spherical-shaped bubble. Here, $\zeta = Z0 - Zcm2$. Here, Z and Zcm represent bubble initial and translation velocity.

Case Study 1

Consider a spherical-shaped acoustic bubble as shown in Figure 3.1 and perform its dynamics studies.

Solution:

Let the spherical bubble radius be "R".

$$\text{Surface area of a spherical particle} = 4\pi R^2 \tag{3.5}$$

So, according to the definition of surface tension,

$$\text{Surface energy} = 4\pi R^2 \sigma \tag{3.6}$$

Let us assume that the bubble is expanding, and the work required to expand the bubble from radius "R" to "$R + dR$" is given in Equation (3.7). Here, the term "dR" representing the differential element of the distance traveled by the bubble.

$$\text{Surface area of a spherical particle after}$$

$$\text{expansion} = 4\pi (R + dR)^2 = 4\pi R^2 + 8\pi\sigma R \cdot dR + 4\pi (dR)^2 \tag{3.7}$$

After eliminating (dR),[2] as it is a smaller value, the above equation can be reduced to,

$$\text{Work required to expand the bubble from radius "}R\text{" to "}R + dR\text{"} = 8\pi\sigma R \cdot dR \tag{3.8}$$

$$\text{Force required to expand the bubble} = \frac{\text{Work}}{\text{Distance travelled}} = \frac{8\pi\sigma R \cdot dR}{dR} = 8\pi\sigma R \tag{3.9}$$

Balance between force inside and outside bubble $= 4\pi R^2 P_{\text{in}} = 4\pi R^2 P_B + 8\pi\sigma R$.

where P_{in} and P_B are representing the pressure inside the bubble and liquid pressure at the bubble wall, respectively,

$$P_{\text{in}} = P_B + \frac{2\sigma}{R} \tag{3.10}$$

where $\dfrac{2\sigma}{R}$ is called as *Laplace pressure*.

According to Equation (3.8), the pressure inside a bubble is higher than the liquid pressure at the bubble wall by the Laplace pressure. In the case of pure water, the Laplace pressure is only a function of bubble radius (R).

Moreover,

$$\text{Laplace pressure} = 1.5 \text{ bar for } R = 1 \text{ μm}$$

$$\text{Laplace pressure} = 15 \text{ bar for } R = 0.1 \text{ μm}$$

Rectified diffusion refers to the process of mass transfer that occurs within a boundary layer that is in motion. The growth occurs due to the uneven transfer of mass between the air and liquid during the rarefaction and compression phases of the sound wave cycle. Gas diffusion takes place over the bubble interface, with a greater amount of gas entering during expansion compared to the amount exiting during contraction. The bubble undergoes growth as gas gathers within it. During the process of compression, the bubble releases gas because the pressure inside the bubble

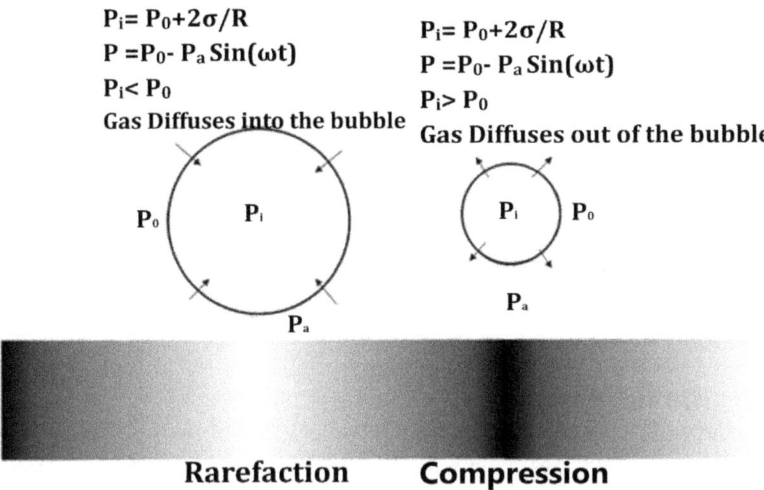

FIGURE 3.2   The schematic image illustrates the minimum and maximum pressure cycles that occur during the rarefaction and compression zones of acoustics.[3]

increases. Gas permeates the bubble when its pressure decreases during expansion as shown in Figure 3.2.

3.3   ACOUSTIC RADIATION FORCES

The study of acoustic bubble dynamics, particularly in fields such as medical ultrasound, acoustic tweezers, and sonochemistry, necessitates the consideration of the substantial impact of acoustic radiation forces. These forces occur as a result of the interaction between the acoustic field and the bubbles, resulting in phenomena such as bubble motion, oscillation, and aggregation.

Principles Governing the Phenomenon of Acoustic Radiation Forces
The acoustic radiation force acting on a bubble can be comprehended by considering two primary constituents:

- The primary Bjerknes force is exerted on each individual bubble as a result of the pressure gradient in the acoustic field.
- The secondary Bjerknes force is a result of the acoustic interactions between pairs of bubbles.

Primary Bjerknes Force
The primary Bjerknes (F_{B1}) force exerts its influence on an individual bubble within an acoustic environment that is not uniform. The expression is defined as:

$$F_{B1} = -V \, \nabla \left(\frac{1}{2}\rho v^2 + p \right)$$

(3.11)

where V is the volume of the bubble, ρ is the density of the liquid, v is the particle velocity of the fluid, p is the acoustic pressure.

The primary Bjerknes force for a spherical bubble in a standing wave field can be estimated as:

$$F_{B1} = 4\pi R^3 \left(\frac{\rho \omega^2 A}{2} \right) \sin(2kx) \qquad (3.12)$$

where R is the bubble radius, ω is the angular frequency of the acoustic wave, "A" is the amplitude of the acoustic pressure, "k" is the wavenumber, "x" is the position in the standing wave field.

Secondary Bjerknes Force

The secondary Bjerknes force arises from the interaction of two oscillating bubbles as a result of the radiation pressure they exert on each other. The expression is defined as:

$$F_{B2} = \frac{3}{4} \pi \rho \left(\frac{R_1^3 R_2^3}{d^2} \right) \left(\frac{\dot{R}_1 \dot{R}_2}{d} \right) e_d \qquad (3.13)$$

where R_1 and R_2 are the radii of the two bubbles, $\dot{R}_1$ and $\dot{R}_2$ are the radial velocities of the bubbles, d is the distance between the centers of the bubbles, e_d is the unit vector along the line connecting the centers of the two bubbles.

3.3.1 APPLICATIONS

Medical ultrasound: Acoustic radiation pressures are employed to manage micro-bubbles in order to provide targeted medicine delivery and improve image quality.

Acoustic tweezers: They utilize the forces to achieve accurate manipulation of minute particles and bubbles within microfluidic devices.

Sonochemistry: A field that studies how forces affect the behavior of bubbles, leading to an enhancement of chemical reactions in solutions.

3.4 YIELDING CRITERIA AND IMPACT ON BUBBLE DYNAMICS

Acoustic bubbles, or cavitation bubbles, are tiny bubbles filled with gas in a liquid that vibrate due to an acoustic field. Understanding the yielding criteria and their impact on bubble dynamics is essential for understanding and directing many applications, including medical ultrasound, sonochemistry, and undersea acoustics. Here is an in-depth examination of these elements:

3.4.1 CRITERIA FOR YIELDING ACOUSTIC BUBBLES

3.4.1.1 Threshold for Acoustic Pressure

- *Cavitation* is initiated when the acoustic wave's negative pressure exceeds the liquid's tensile strength. This phenomenon occurs when the amplitude of the sound pressure is beyond a specific threshold.

- *Bubble expansion and contraction*: Once cavitation initiates, the bubble will undergo enlargement during the phase of negative pressure and contraction during the phase of positive pressure of the acoustic wave. The size of the bubble at its maximum and the strength of its collapse are contingent upon the amplitude of the acoustic pressure.

3.4.1.2 Acoustic Field Frequency

- Resonance occurs when bubbles vibrate at particular frequencies that are influenced by their size and the characteristics of the surrounding substance. When bubbles are in resonance, they can oscillate with greater amplitudes, resulting in more noticeable dynamic behaviors.
- Low-frequency acoustic fields generally result in inertial cavitation, which is characterized by the forceful collapse of bubbles. High-frequency fields typically generate non-inertial cavitation, characterized by more steady oscillations.

3.4.1.3 Characteristics of Liquids

- Viscosity and surface tension have the ability to inhibit the production and expansion of bubbles. This means that larger sonic pressures are needed to induce cavitation when viscosity and surface tension are increased.
- *Gas content*: The existence of gases dissolved in the liquid can aid in the formation of bubbles and impact the process of bubble expansion and contraction.

3.4.1.4 Size and Distribution of Bubble Nuclei

- *Pre-existing nuclei*: Minute gas nuclei present in the liquid serve as predecessors for the development of bubbles.
- The size distribution and concentration of these nuclei influence the minimum level at which cavitation occurs and the subsequent behavior of the resultant bubbles as shown in Figure 3.3.

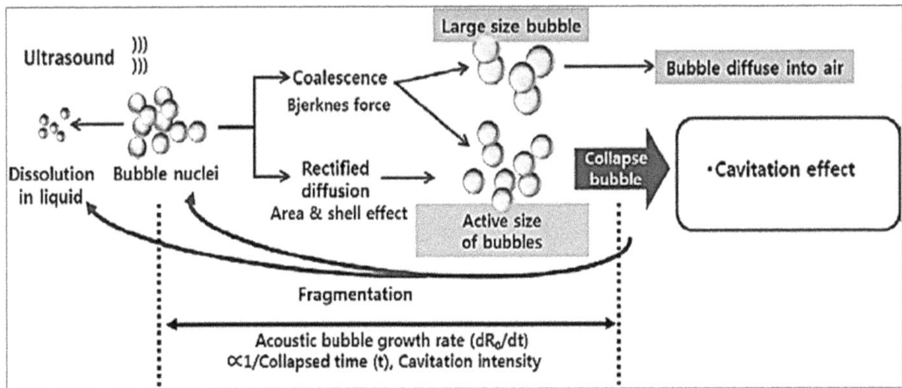

FIGURE 3.3 Diagram illustrating the process of bubble expansion through a corrected diffusion pathway and a coalescence pathway.[4]

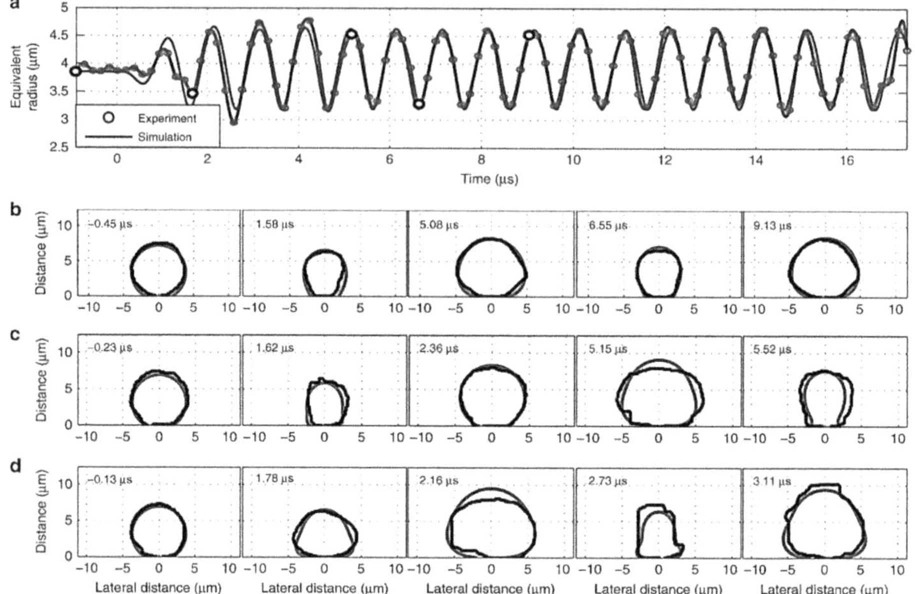

FIGURE 3.4 Nonspherical microbubble oscillations. (a) Simulated and experimental radius–time curve of a 3.9-μm radius microbubble exposed to 166 kPa ultrasound burst at 1 MHz and its simulation contours mentioned in (b). (c) and (d) show the simulation contours of a bubble exposed to 249 and 331 kPa, of ultrasound burst, respectively.[4]

3.5 IMPACT ON BUBBLE DYNAMICS

3.5.1 Oscillation Behavior

- Linear oscillations occur when bubbles experience tiny and stable fluctuations in size around a state of equilibrium, particularly at low acoustic pressures.
- *Nonlinear oscillations*: When subjected to higher pressures, bubbles display nonlinear characteristics, such as oscillations in their shape (e.g., becoming nonspherical) and intricate radial dynamics as shown in Figure 3.4.
- Bubble collapse refers to the phenomenon where a bubble, often formed in a liquid or gas, suddenly collapses or implodes due to various factors.

3.5.2 Bubble Collapse

The collapse of bubbles under the influence of an acoustic field can result in important physical phenomena, including inertial cavitation and microstreaming as shown in Figure 3.5. Comprehending these phenomena is crucial for their utilization in medical ultrasound, sonochemistry, and diverse engineering disciplines. Below is an in-depth analysis of inertial cavitation and microstreaming:

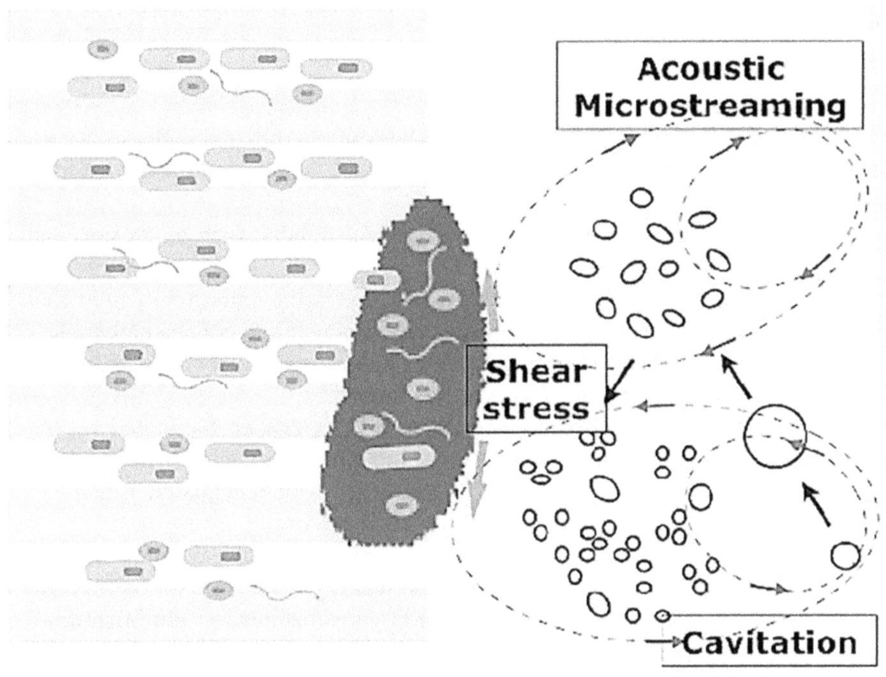

FIGURE 3.5   Illustration depicting the mechanism of inertial cavitation and microstreaming.[5]

- *Inertial cavitation*: Inertial cavitation is defined as the rapid collapse of a bubble, resulting in the generation of elevated temperatures and pressures within the bubble. This phenomenon can generate shock waves, emit light (known as sonoluminescence), and trigger chemical reactions.

  i **Mechanism**

  High acoustic pressure: It refers to the situation where a bubble, when subjected to an acoustic field, expands significantly during the rarefaction phase and subsequently collapses forcefully during the compression phase. This phenomenon is known as inertial cavitation.

  Typically, achieving large bubble expansion and subsequent collapse necessitates the use of high acoustic pressures.

  Bubble dynamics: In the rarefaction phase, the bubble undergoes expansion due to the decrease in pressure of the surrounding liquid. As the pressure rises during the compression phase, the bubble undergoes a fast collapse.

  The collapse can be almost adiabatic, resulting in highly intense conditions within the bubble.

  Extreme conditions: Under extreme conditions, the quick collapse of the bubble results in the generation of extremely high temperatures, reaching several thousand Kelvin, and pressures, reaching hundreds of atmospheres.

The severe conditions can lead to the separation of water molecules, resulting in the formation of free radicals and other highly reactive substances.

ii. **Implications**

Shock waves: Shock waves can be generated by the collapse, which then travel through the liquid and result in concentrated impacts of high pressure.

These shock waves cause mechanical effects, such as erosion and pitting, on surfaces.

Sonoluminescence: Sonoluminescence is the emission of light that occurs as a result of the intense conditions within a collapsing bubble.

The phenomenon is perceived as momentary bursts of illumination resulting from the implosion of the bubble.

Chemical reactions: Chemical reactions can be facilitated by the application of high temperatures and pressures, which makes inertial cavitation a valuable tool in the field of sonochemistry. It helps to increase reaction rates and alter reaction routes.

Biological effects: In medical contexts, inertial cavitation has the ability to destroy cellular structures, making it valuable for therapeutic objectives such as tissue ablation or targeted medication administration.

Nevertheless, if not meticulously regulated, it can also result in undesired harm to adjacent tissues.

- *Microstreaming*: It refers to the phenomenon where oscillating bubbles cause a concentrated movement of fluid, resulting in improved mixing and transfer of mass in the surrounding liquid.

Mechanism

1. **Oscillating Bubbles:**
   - Microstreaming is the phenomenon where bubbles oscillate in a steady manner due to an acoustic field, usually at lower acoustic pressures than those required for inertial cavitation.
   - The oscillations of the bubble's surface create localized fluid flows around the bubble.[6]

2. **Steady Streaming:**
   - Microstreaming refers to the time-averaged, constant flows that are generated in the surrounding liquid due to the oscillatory motion of the bubble.
   - The flow patterns are contingent upon the frequency of the acoustic field, the dimensions of the bubble, and the viscosity of the liquid.

3. **Implications:**

   Enhanced mixing: Microstreaming causes concentrated blending within the liquid, which improves the rates of mass transfer and diffusion. Enhanced mixing can be advantageous in processes such as sonochemistry, as it can result in more efficient reactions.[7]

   Biological applications: Microstreaming in medical ultrasound can increase the permeability of cell membranes, making it easier to administer drugs or transfer genes. Additionally, it can aid in the dissolution of blood clots or enhance tissue perfusion by augmenting fluid circulation.

Surface cleaning: Microstreaming is a technique that can be employed in cleaning processes to effectively eliminate impurities from surfaces by utilizing the induced flow of fluid.

Microfluidics: Microfluidics refers to the study and manipulation of fluids at the microscale level. Microstreaming is a technique employed in microfluidic devices to manage fluids and particles on a tiny level. It allows for accurate control of mixing and transport processes.

3.5.3   ACOUSTIC EMISSIONS

3.5.3.1   Harmonic Generation

Harmonic generation refers to the production of harmonics and subharmonics of the driving acoustic frequency through the nonlinear oscillations of bubbles. These generated frequencies can be monitored and utilized for diagnostic reasons.[8]

3.5.3.2   Noise Production

The occurrence of forceful bubble collapses generates a wide range of noise frequencies, which has the potential to disrupt underwater communication and sonar systems.

3.5.3.3   Thermal and Fluid Dynamics

Oscillating bubbles boost mass transfer rates by enhancing surface area and mixing effects, resulting in enhanced diffusion.

3.5.3.4   Thermal Effects

The collapse of bubbles produces concentrated heat, which can impact the speeds of chemical reactions and the biological tissue in medical contexts.[9]

3.5.4   APPLICATIONS

3.5.4.1   Ultrasound in the Field of Medicine

- Drug delivery can be improved by utilizing controlled cavitation, which temporarily enhances the permeability of cell membranes.
- Tissue ablation is achieved by the use of high-intensity focused ultrasound, which utilizes cavitation as a noninvasive method for destroying cancerous tissue.

3.5.4.2   Sonochemistry

- Sonochemistry is the study of chemical reactions that are induced or enhanced by sound waves. Cavitation bubbles enhance chemical reactions by creating severe conditions and improving the transfer of mass.
- Nanomaterial cavitation can be utilized for the production of nanoparticles and other sophisticated materials through the process of synthesis.[10]

3.5.4.3 Subaquatic Acoustics

- *Sonar performance*: The presence of cavitation noise can disrupt sonar systems, hence impacting their ability to detect and communicate.
- *Marine engineering*: Cavitation erosion is a significant challenge for ship propellers and hydraulic systems, necessitating the use of materials and designs that effectively reduce cavitation-induced damage.

3.6 SHAPE OSCILLATIONS

Although early investigations frequently focus on spherical bubbles, real bubbles can have intricate form oscillations. Nonspherical oscillations arise when the bubble's symmetry is disrupted by variables such as intense sound waves or interactions with adjacent bubbles.[11]

3.6.1 MODE SHAPES AND FREQUENCIES

Shape oscillations can be categorized by their mode shapes, which are described by spherical harmonics. The simplest nonspherical oscillations involve the $l = 2$ mode, known as quadrupole oscillations. Higher modes ($l = 3, 4, ...$) correspond to more complex shapes. The frequency of these shape oscillations depends on the mode number "l" and the equilibrium radius R_0 of the bubble:

$$\text{Angular frequency at } l\text{th mode } (\omega_l) = \sqrt{\frac{l(l-1)(l+1)(l+2)\sigma}{\rho R_0{}^3}} \tag{3.14}$$

where ω_l is the angular frequency of the lth mode, σ is the surface tension of the liquid, ρ is the density of the liquid, R_0 is the equilibrium radius of the bubble.

3.7 SUMMARY

- The investigation of bubble dynamics and form oscillations encompasses sophisticated principles of fluid mechanics and acoustic theory, with practical ramifications for improving the process of mixing, localized heating, and targeted delivery systems.
- The Rayleigh-Plesset equation forms the cornerstone of theoretical frameworks for understanding the oscillatory behavior of acoustic bubbles.
- To control the shape oscillations of bubbles in real-life situations, one must manipulate the parameters of the acoustic field, such as frequency and amplitude.
- Various factors, such as the existence of contaminants, adjacent bubbles, and boundary conditions (such as being close to solid surfaces), also affect the behavior of bubbles.
- Comprehending and managing the oscillations are essential for maximizing efficiency in diverse industrial and medical applications.

REFERENCES

1. Besant, W. H. (1859). *A treatise on Hydrostatics and Hydrodynamics*. Deighton: Bell.
2. Riccardi, G., & De Bernardis, E. (2016). Dynamics and acoustics of a spherical bubble rising under gravity in an inviscid liquid. *The Journal of the Acoustical Society of America*, 140(3), 1488–1497. https://doi.org/10.1121/1.4962160
3. Leong, T., Ashokkumar, M., & Kentish, S. (2016). The growth of bubbles in an acoustic field by rectified diffusion. In M. Ashokkumar (Ed.), *Handbook of Ultrasonics and Sonochemistry* (pp. 69–98). https://doi.org/10.1007/978-981-287-278-4_74
4. Kang, B. K., Kim, M. S., & Park, J. G. (2014). Effect of dissolved gases in water on acoustic cavitation and bubble growth rate in 0.83 MHz megasonic of interest to wafer cleaning. *Ultrasonics Sonochemistry*, 21(4), 1496–1503.
5. Lajoinie, G., Luan, Y., Gelderblom, E. et al. (2018). Non-spherical oscillations drive the ultrasound-mediated release from targeted microbubbles. *Communications Physics,* 1, 22. https://doi.org/10.1038/s42005-018-0020-9
6. Kim, S. G. (2016). Infection and pulp regeneration. *Dentistry Journal*, 4(1), 4.
7. Sonawane, S. S., Charde, S. J., Malika, M., & Thakur, P. (2022). Artificial neural network model for prediction of viscoelastic behaviour of polycarbonate composites. *Journal of Applied Research and Technology*, 20(2), 188–202.
8. Malika, M., & Sonawane, S. S. (2022). The sono-photocatalytic performance of a Fe_2O_3 coated TiO_2 based hybrid nanofluid under visible light via RSM. *Colloids and Surfaces A: Physicochemical and Engineering Aspects*, 641, 128545.
9. Malika, M., & Sonawane, S. S. (2022). MSG extraction using silicon carbide-based emulsion nanofluid membrane: Desirability and RSM optimisation. *Colloids and Surfaces A: Physicochemical and Engineering Aspects*, 651, 129594.
10. Malika, M., Jhadav, P. G., Parate, V. R., & Sonawane, S. S. (2023). Synthesis of magnetite nanoparticle from potato peel extract: Its nanofluid applications and life cycle analysis. *Chemical Papers*, 77(2), 1081–1094.
11. Malika, M., Pargaonkar, A., & Sonawane, S. S. (2023). Application of emulsion nanofluid membrane for the removal of methylene blue dye: Stability study. *Chemical Papers*, 77(7), 3967–3977.

4 Introduction to Nanofluids

4.1 INTRODUCTION

Various methods, such as vanes, fins, radiators, and forced air/fluid through cooling channels, are employed to cool or regulate the temperature in these devices, despite their high cost. Certain equipment and systems utilize cost-effective traditional heat transfer fluids to enhance heat dissipation, including water (deionized water), ethylene glycol (EG), oils, and other lubricants. Nevertheless, these fluids have a drawback in terms of their low thermal conductivity. For example, water is approximately three orders of magnitude less conductive than copper (Table 4.1). However, the lack of thermal conductivity in these typical fluids is offset by their capacity to flow.

4.2 ORIGIN OF NANOFLUIDS

The emergence of nanofluids can be traced back to developments in both theoretical and experimental studies in thermal conductivity and fluid mechanics. Below is a comprehensive chronology and elucidation of the significant events that have led to the advancement of nanofluids:

4.2.1 EARLY THEORETICAL FOUNDATIONS

James Clerk Maxwell (1873): Maxwell's research on the thermal conductivity of composite materials established the theoretical foundation for comprehending the influence of suspended particles on the overall thermal conductivity of a fluid. The model (Equation 4.1) focused on particles that were on a micron scale and offered a formula to calculate the overall thermal conductivity of a mixture. This calculation was based on the characteristics and volume fractions of the components in the mixture.

$$k_{\text{eff}} = k_{\text{Base fluid}} \left(\frac{k_{\text{Nanoparticle}} + 2k_{\text{Base fluid}} + 2\varphi\left(k_{\text{Nanoparticle}} - k_{\text{Base fluid}}\right)}{k_{\text{Nanoparticle}} + 2k_{\text{Base fluid}} - \varphi\left(k_{\text{Nanoparticle}} - k_{\text{Base fluid}}\right)} \right) \tag{4.1}$$

where

k_{eff} is the effective thermal conductivity of nanofluid
φ is the solid volume fraction of nanofluids

Nevertheless, when employed in the context of nanofluids, this model is subject to various constraints.

DOI: 10.1201/9781003594949-4

TABLE 4.1
Thermal Conductivity (W/mK) of Different Materials at 25°C Room Temperature

Material	Thermal conductivity (W/mK)
Fluids	
Deionized water	0.603
Hexane	0.126
Glycerol	0.285
Engine oil	0.145
Ethanol	0.172
Toluene	0.133
Ethylene glycol	0.252
Metallic solids	
Copper	401
Aluminum	257
Silver	429
Gold	345
Nickel	158
Iron	80
Non-metallic solids	
Carbon nanotubes	2,000–6,000
Diamond	2,000
Silica	148
SiC	270
Al_2O_3	40
CuO	76
Fe_2O_3	7
ZnO	29
TiO_2	25

4.2.1.1 Limitations of Maxwell's Theory

1. Contemplation of Particle Size

   Maxwell's model: The model was initially designed for particles of a micron size, and it does not consider the specific phenomena that occur at the nanoscale, such as the ratio of surface area to volume and quantum effects.

   Nanofluids consist of nanoparticles that possess much greater surface area to volume ratios. These ratios have a substantial impact on thermal conductivity, a phenomenon that is not accounted for in Maxwell's theory.

2. Effects and Interactions on the Surface

   Maxwell's model assumes that particles are uniformly dispersed and exhibit negligible molecular-level interactions with the fluid.

Nanofluids are characterized by the strong interaction between nanoparticles and the fluid molecules in their vicinity. This interaction gives rise to intricate phenomena, such as the creation of interfacial layers. The thermal conductivity of nanofluids can be either enhanced or reduced, depending on the specific combination of fluid and particle materials.

3. Brownian Motion

Brownian motion refers to the random movement of particles suspended in a fluid due to collisions with molecules in the fluid.

Maxwell's model fails to account for the influence of Brownian motion, which becomes substantial when dealing with nanoparticles. Brownian motion can increase thermal conductivity by enabling supplementary mechanisms for energy transmission.

Nanofluids have improved thermal transport qualities due to the Brownian motion of nanoparticles, a phenomenon that is not accounted for in Maxwell's model.

4. Aggregation and Stability

Maxwell's model assumes that the particles are stable and evenly distributed. It fails to consider particle aggregation, which is more likely to happen at the nanoscale because of van der Waals forces and other interactions.

The thermal conductivity and stability of nanofluids can be greatly influenced by the aggregation of nanoparticles. Aggregates can modify the flow properties and decrease the usable surface area for heat transmission.

5. Influence of Nanoparticle Shape and Morphology

Maxwell's model assumes the simplification of spherical particles, hence restricting its applicability to particles of different forms.

Nanofluids consist of nanoparticles of different forms, such as rods, tubes, and platelets, which have an impact on their thermal conductivity capabilities. Nanoparticles that are not spherical generally have varying thermal conductivities because of the presence of anisotropic heat conduction channels.

6. Interfacial Thermal Resistance

Maxwell's model does not account for interfacial thermal resistance, which refers to the resistance to heat transfer at the interface between the nanoparticles and the base fluid.

The overall thermal conductivity of nanofluids can be greatly affected by the existence of interfacial thermal resistance. The resistance can be especially pronounced for nanoparticles that have limited compatibility with the base fluid.

7. Non-Newtonian Phenomena

Maxwell's model assumes that the base fluid behaves according to Newtonian principles, with a constant viscosity that does not change with the rate of shear.

Nanofluids: Certain nanofluids display non-Newtonian characteristics, wherein the viscosity varies with the rate of shear, hence impacting the heat transfer capabilities. Maxwell's model does not account for this phenomenon.

4.2.2 Development of Colloids and Suspensions

During the early to mid-20th century, scientists conducted further research on colloids and suspensions, specifically examining the characteristics and behavior of bigger particles (measuring in microns) that are dispersed in fluids. The main focus of these investigations was on stability, viscosity, and thermal characteristics. However, the improvements in thermal conductivity were quite minor in comparison with the significant advances shown in nanofluids.

4.2.3 Emergence of Nanotechnology

In the 1980s, the emergence of nanotechnology allowed for the creation and control of particles that are smaller than 100 nanometers in size, known as nanoparticles. This technological breakthrough has created new opportunities for improving material characteristics, such as thermal conductivity.

4.2.4 Introduction to Nanofluids

Choi and Eastman[1] published a work in 1995. Choi and Eastman, researchers at Argonne National Laboratory, proposed the notion of nanofluids. The researchers suggested and proved that the addition of nanoparticles to base fluids might greatly improve their thermal conductivity. This discovery was significant since the measured improvements were significantly larger than what classical theories, such as Maxwell's theory, predicted for particles that are one millionth of a meter in size.

4.2.4.1 Significant Contributions Made by Choi and Eastman

Improved heat transfer: Their investigations showed that even a minimal proportion of nanoparticles could result in significant enhancements in thermal conductivity. The unique qualities of nanoparticles, such as their huge surface area and quantum effects, are responsible for this phenomenon, which is not observed in larger particles.

4.2.4.2 Emerging Areas of Study

The advent of nanofluids has prompted extensive research into comprehending the principles underlying the improved thermal characteristics and investigating different categories of nanoparticles (such as metals, oxides, and carbon nanotubes) and base fluids.

4.2.5 Subsequent Developments

1995 to present: Subsequent to the groundbreaking research accomplished by Choi and Eastman, there has been a significant amount of investigation aimed at enhancing our comprehension and maximizing the efficiency of nanofluids. Research has mostly concentrated on many characteristics like the morphology, dimensions, abundance, surface modification, and durability of the nanofluids.

The origin of nanofluids can be summarized as follows:

- In 1873, Maxwell laid the theoretical foundation for comprehending the thermal conductivity of particle suspensions.
- Throughout the 20th century, research on colloids and suspensions continued, albeit with limited improvements in thermal properties.
- However, in the 1980s, advancements in nanotechnology facilitated the production of nanoparticles.
- It was in 1995 that Choi and Eastman introduced nanofluids, demonstrating significant enhancements in thermal conductivity through the use of nanoparticles. Since then, ongoing research has further deepened our understanding of nanofluids and expanded their range of applications.

4.2.6 GLOBAL DISTRIBUTION OF PUBLICATIONS ON NANOFLUIDS

Over the past decade, there has been a notable increase in the amount of research conducted on nanofluids. This has entailed performing multiple experiments and theoretical investigations to examine different facets of nanofluids. A search was conducted in the Scopus database to locate scholarly research papers on the subject of "Nanofluids." The search and subsequent analysis cover the time period from 2010 to 2022, as shown in Figure 4.1. A collective of 124 countries and territories

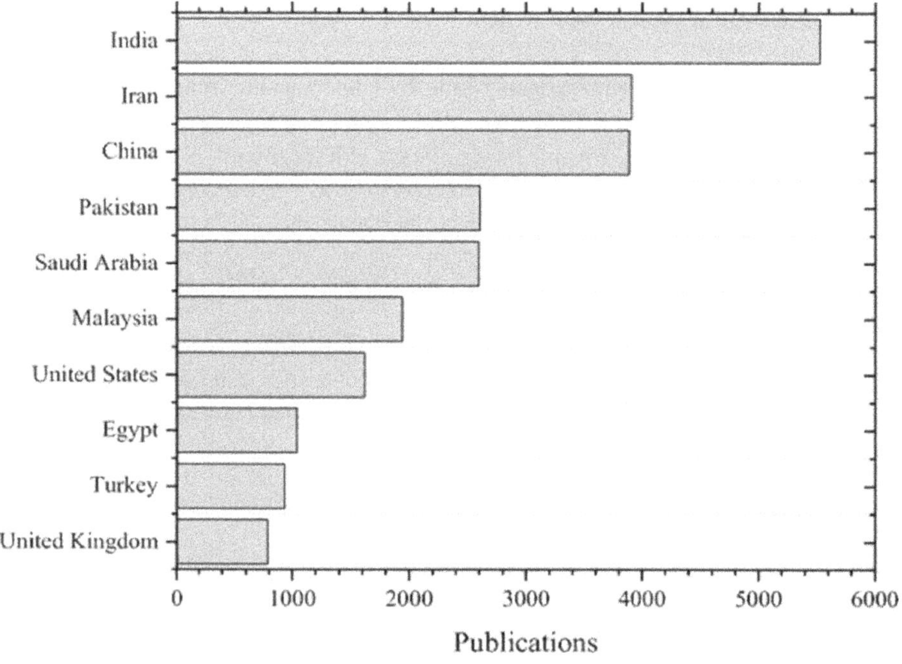

FIGURE 4.1 Regional distribution of nanofluid publications from 2010 to 2022.[2]

contributed articles to the scene. India has attained the top position in terms of nano-fluid publications, having a grand total of 5,524 papers. Similarly, researchers from other countries made noteworthy contributions.

4.3 FUNDAMENTALS OF NANOFLUIDS

4.3.1 DEFINITION AND TYPES OF NANOFLUIDS

Nanofluids are defined as suspensions of nanoparticles (metallic, non-metallic, or polymeric) in a base fluid. Nanofluids are precisely designed mixtures of nanoparticles (1–100 nm and <1 Vol.%) with non-dissolving base fluids like water, EG, or oils. Nanoparticles in these fluids confer distinct characteristics, such as heightened thermal conductivity (Figure 4.2), viscosity, and heat transport capacities, setting them apart from regular fluids. The term "nanofluids" was initially introduced by Choi and Eastman[1] in 1995 (Figure 4.3). Subsequently, significant research efforts have been undertaken to comprehend and enhance their characteristics for diverse applications, such as heat exchangers, cooling systems, and biomedical applications. Moreover, the thermophysical properties (also known as the effective parameters) of the colloid are heavily influenced by the following factors:

- Nanoparticles
- The factors to consider while analyzing a material include its type, size, shape, attraction, or extraction properties with the hosting basefluid

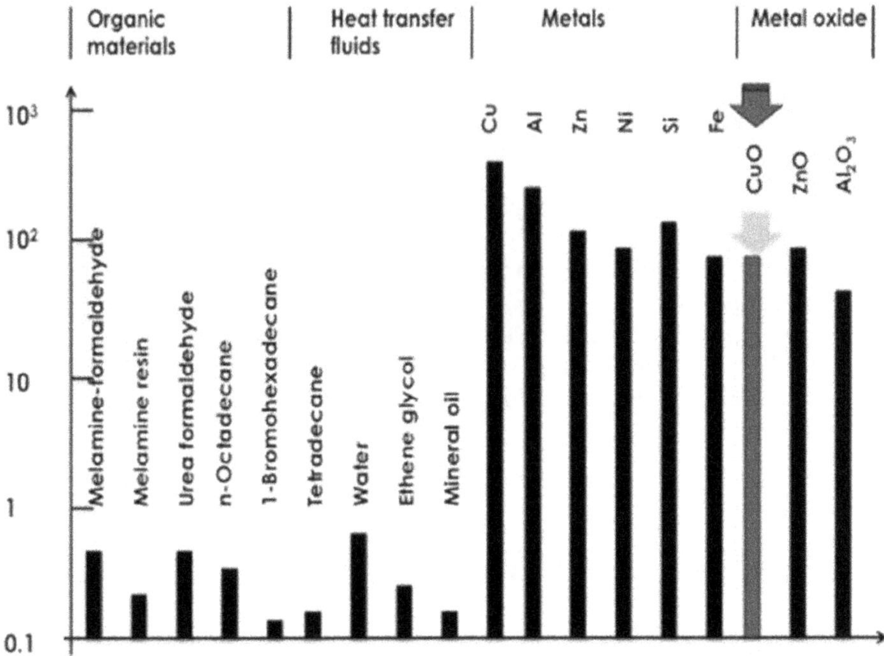

FIGURE 4.2 Thermal conductivity comparison of different materials.[2]

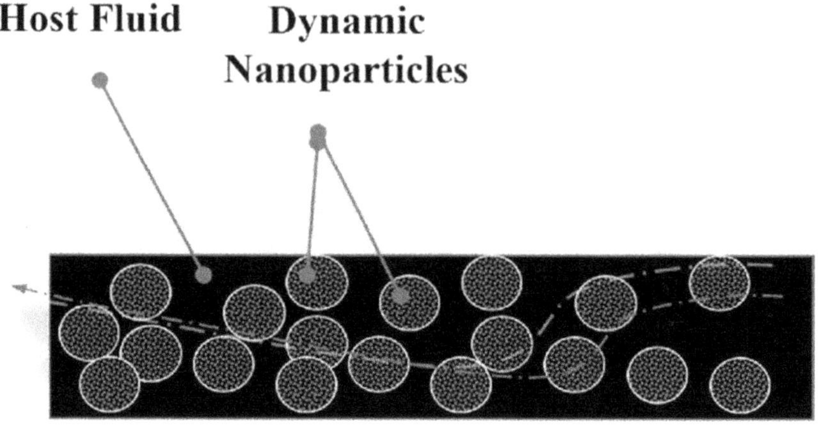

FIGURE 4.3   The universal configuration of a nanofluid.[2]

molecules, volumetric concentration, density, specific heat capacity, and thermal conductivity.

- Base fluids

Base fluids can be characterized by several properties, including their kind, temperature, pH value, molecular attraction, or extraction behavior toward dispersed particles, density, specific heat capacity, viscosity, and thermal conductivity.

- Preparation methodology includes a single-step or a two-step approach.
- Chemical or physical dispersion(s) may be present, if introduced.
- Stability of nanofluids in terms of dispersion, kinetics, and chemical properties over both long and short periods.

They are classified based on the type of nanoparticles used:

- *Metallic nanofluids*: Contain metal nanoparticles like gold, silver, copper, or aluminum.
- *Oxide nanofluids*: Include oxide nanoparticles such as alumina (Al_2O_3), titania (TiO_2), and silica (SiO_2).
- *Carbon-based nanofluids*: Comprise carbon nanotubes (CNTs), graphene, and fullerenes.
- *Polymeric nanofluids*: Involve polymer nanoparticles suspended in the base fluid.

4.4 PROPERTIES OF NANOFLUIDS

The enhanced properties of nanofluids are primarily attributed to the high surface area to volume ratio of the nanoparticles and their interaction with the base fluid. Figure 4.4 shows the important thermophysical properties of nanofluids over conventional base fluids. Key properties include:

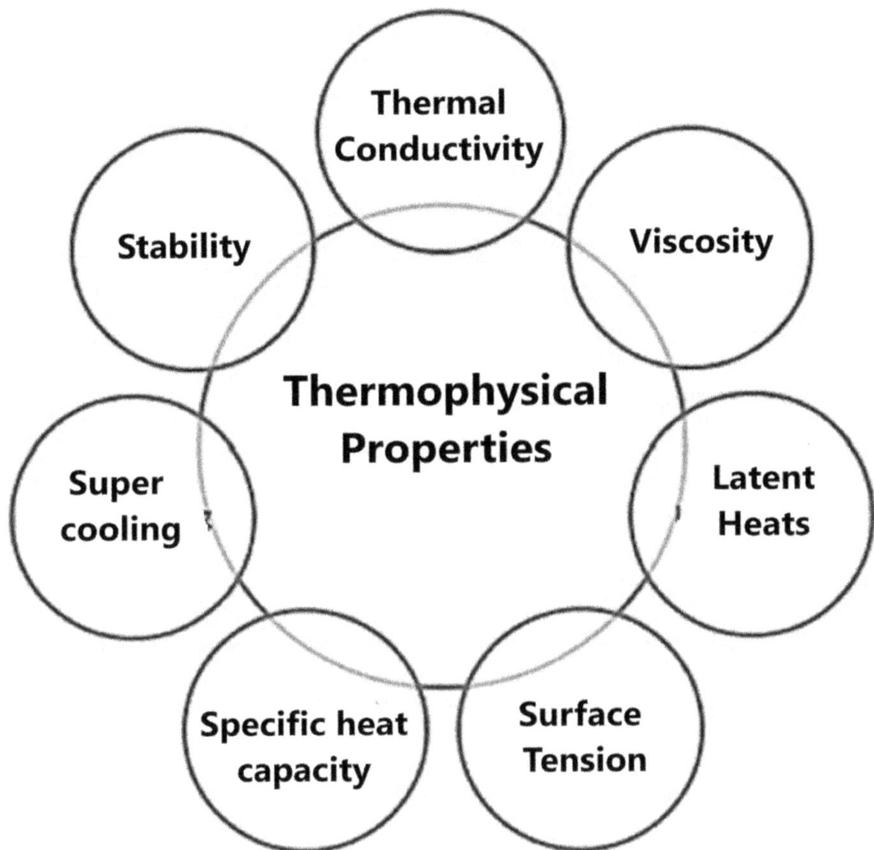

FIGURE 4.4 Key thermophysical properties of nanofluids.[3]

4.4.1 Increased Heat Transfer Ability

An outstanding characteristic of nanofluids is their heightened heat conductivity in comparison with base fluids, such as water, EG, or oil. The incorporation of nanoparticles greatly enhances the thermal conductivity of the fluid.

4.4.2 Enhanced Thermal Conductivity

Nanofluids frequently have elevated convective heat transfer coefficients. This feature is essential for applications such as cooling systems, where efficient heat dissipation is necessary.

4.4.3 Stability and Dispersion

The stability of nanoparticles in the base fluid is crucial for maintaining consistent performance of nanofluids. Ensuring stable dispersion prevents the nanoparticles from clumping together or settling, which would otherwise reduce their thermal characteristics.

4.4.4   Viscosity

The viscosity of nanofluids may surpass that of the base fluid, contingent upon the concentration and nature of the nanoparticles employed. This can impact the amount of power needed to pump fluids in flow systems and should be tuned for specific uses.

4.4.5   Specific Heat Capacity

The kind and concentration of nanoparticles can have an impact on the specific heat capacity of nanofluids. This attribute quantifies the fluid's capacity to retain thermal energy.

4.4.6   Rheological Properties

Nanofluids have distinct rheological properties compared to ordinary fluids. Comprehending the rheological characteristics is crucial for forecasting and controlling the flow behavior in diverse applications.

4.4.7   Thermal Diffusivity

An additional important property is thermal diffusivity, which is a combination of thermal conductivity, density, and specific heat. Nanofluids typically exhibit increased thermal diffusivity, resulting in accelerated heat propagation within the fluid.

4.4.8   Electrical Conductivity

Nanofluids can display modified electrical conductivity, which can have either advantageous or disadvantageous effects depending on the specific nanoparticles employed and the intended usage.

4.4.9   Optical Properties

Nanofluids exhibit distinctive optical characteristics, such as enhanced light absorption or light scattering. These properties render them valuable in fields such as solar energy collection and photothermal therapy.

4.4.10   Corrosion and Wear Resistance

Nanofluids can offer superior corrosion and wear resistance in comparison with traditional fluids, hence increasing the longevity of the systems in which they are employed.

4.4.11   Magnetic Characteristics

Nanofluids that contain magnetic nanoparticles possess magnetic properties, which render them valuable for applications such as magnetic resonance imaging (MRI) and targeted drug administration.

4.4.12 Surface Tension

Nanofluids can exhibit surface tension that deviates from that of the base fluid, hence influencing capillary action and boiling heat transfer mechanisms.

Nanofluids include features that make them extremely adaptable and desirable for a wide range of industrial and technological uses. These include the cooling of electronic devices, automobile cooling systems, solar collectors, and biological applications. Subsequent chapters will thoroughly examine these enhanced characteristics of nanofluids.

4.5 PREPARATION AND CHARACTERIZATION OF NANOFLUIDS

Synthesis of nanofluids involves several crucial steps to ensure stability, uniformity, and desired properties. The preparation of nanofluids involves dispersing nanoparticles into a base fluid as shown in Figure 4.5. The main methods include:

- *Two-step method*: Nanoparticles are first synthesized and then dispersed in the base fluid using ultrasonic agitation or high-shear mixing.
- *One-step method*: Simultaneous synthesis and dispersion of nanoparticles in the base fluid, often using techniques like chemical reduction or thermal decomposition.

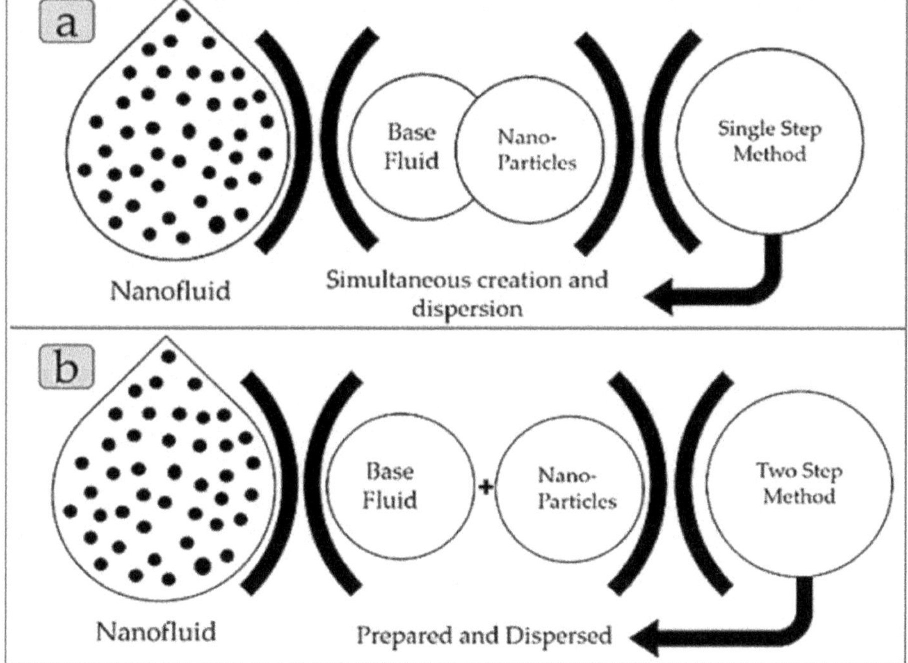

FIGURE 4.5 Nanofluid synthesis methodologies: (a) one-step and (b) two-step methods.[4]

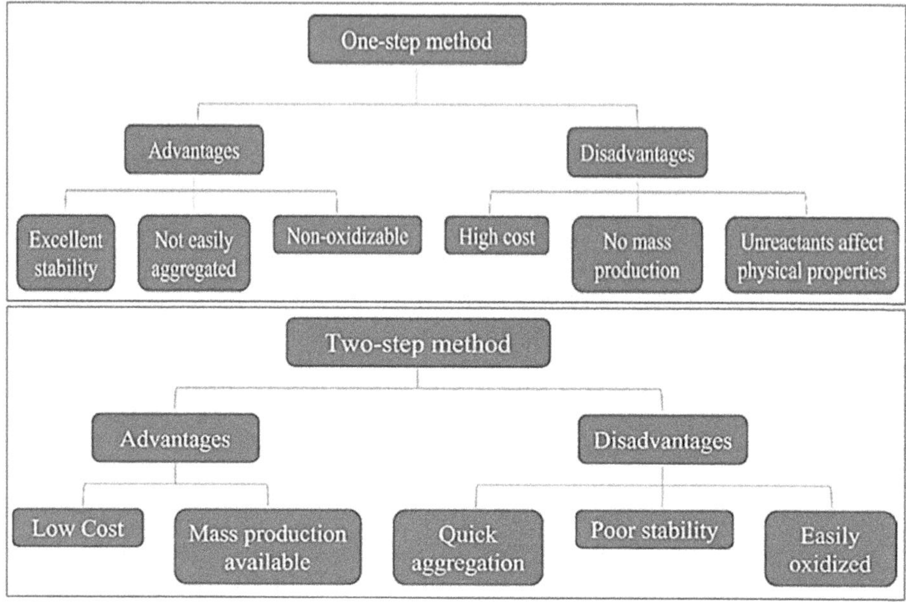

FIGURE 4.6 Advantages and disadvantages of one-step (top) and two-step methods (bottom) of nanofluids.

- *Direct evaporation*: Involves evaporating the base fluid and condensing it in the presence of nanoparticles.
- *Microwave-assisted synthesis*: Uses microwave radiation to synthesize and disperse nanoparticles simultaneously.

Among these methodologies for a small-scale and large-scale production of nanofluids, one-step and two-step methods are utilized, respectively. Figure 4.6 elaborately explains the advantages, disadvantages, and limitations of each method. Where, one-step method, the preparation, and dispersion of nanoparticles are carried out simultaneously. This method aids in resolving concerns pertaining to agglomeration and guarantees enhanced stability of the nanofluid. Several frequently used one-step techniques include:

- *Direct evaporation*: It is a process of evaporating a nanoparticle precursor directly into the base fluid.
- *Chemical reduction*: It is a synthesis of nanoparticles by reducing a metal salt in a base solution.
- Laser ablation is a process where a laser is used to remove material from a target substance, while a base fluid is present. This causes the production and scattering of nanoparticles.

4.6 CHARACTERIZATION TECHNIQUES

Characterizing nanofluids entails evaluating their physical and chemical properties to verify their stability and performance:

- Particle size and distribution can be determined by the use of dynamic light scattering (DLS) or transmission electron microscopy (TEM).
- The Zeta potential is a measurement of the charge on the surface of nanoparticles, which indicates the stability of nanofluids.
- Thermal conductivity can be determined by many methods such as the transient hot-wire method, steady-state approach, or laser flash analysis.
- Viscosity is measured by use rheometers or viscometers.
- *Specific heat capacity*: It is determined through the use of differential scanning calorimetry (DSC).

4.7 STABILITY OF NANOFLUIDS

Ensuring the stability of nanofluids is crucial for their effective application. Stability can be influenced by:

- *pH adjustment*: Adjusting the pH of the suspension to enhance particle dispersion.
- *Surfactants*: Adding surfactants to prevent agglomeration of nanoparticles.
- *Ultrasonication*: Using ultrasonic waves to break down nanoparticle clusters and improve dispersion.

4.8 FACTORS INFLUENCING THERMOPHYSICAL PROPERTIES OF NANOFLUIDS

4.8.1 Particle Size and Shape

The size and shape of nanoparticles significantly affect the thermal conductivity and viscosity of nanofluids. Smaller particles with higher surface area to volume ratio enhance thermal conductivity but may increase viscosity.

4.8.2 Particle Concentration

Higher concentrations of nanoparticles generally lead to better thermal conductivity but can increase viscosity and sedimentation issues. Optimal concentration balances these effects for specific applications.

4.8.3 Base Fluid Properties

The properties of the base fluid, such as viscosity, specific heat, and thermal conductivity, influence the overall performance of the nanofluid. The interaction between the base fluid and nanoparticles is crucial for effective heat transfer.

4.8.4 TEMPERATURE

The temperature of the nanofluid affects its thermophysical properties. As temperature increases, the thermal conductivity typically increases, while viscosity decreases, enhancing flow characteristics.

4.8.5 BROWNIAN MOTION

Brownian motion of nanoparticles contributes to enhanced thermal conductivity by facilitating energy transfer within the fluid. This effect is more pronounced in smaller nanoparticles and higher temperatures.

4.9 THEORETICAL ASPECTS OF NANOFLUIDS CHARACTERISTICS

Effective medium theories are used to predict the thermal conductivity of nanofluids based on the properties of the base fluid and nanoparticles. Common models are explained further.

4.9.1 MAXWELL MODEL

The Maxwell model, formulated by James Clerk Maxwell in 1873, is among the early models used to forecast the effective thermal conductivity of composite materials. It is specifically used for handling diluted solutions of spherical particles. The model implies that the particles are evenly distributed and do not exhibit any interaction between them.

The thermal conductivity of a nanofluid, as determined by the Maxwell model, can be expressed as:

$$k_{\mathrm{Eff}} = k_{\mathrm{BF}} \left(\frac{k_{\mathrm{Np}} + 2k_{\mathrm{BF}} + 2\varphi\left(k_{\mathrm{NP}} - k_{\mathrm{BF}}\right)}{k_{\mathrm{Np}} + 2k_{\mathrm{BF}} - \varphi\left(k_{\mathrm{NP}} - k_{\mathrm{BF}}\right)} \right) \tag{4.2}$$

where

k is the thermal conductivity (W/mK)

Subscript:

     Eff: Effective
     BF: Base fluid
     NP: Nanoparticle
     φ: Solid volume fraction of nanofluids

4.9.1.1 Applications

- Dilute suspensions are particularly suitable for situations when the volume percentage of nanoparticles is low, and the interactions between particles can be considered insignificant.

- Spherical nanoparticles are the most suitable choice for this application due to their inherent assumptions.
- *Basic fluids*: Utilized in situations where the primary fluid is simple and its characteristics are widely understood.

4.9.1.2 Limitations

- *Low volume fractions* are not suitable for accurately representing high volume fractions of nanoparticles due to the lack of consideration for particle interactions.
- *Spherical assumption*: This assumption considers nanoparticles to be spherical, which reduces its accuracy when applied to particles that are not spherical.
- Disregards any interactions between nanoparticles and the influence of Brownian motion.

4.9.2 HAMILTON-CROSSER MODEL CONSIDERS THE SHAPE OF NANOPARTICLES AND THEIR ASPECT RATIO

The Hamilton-Crosser model expands upon the Maxwell model by including the morphology of the dispersed particles. This model has a broader scope and can be utilized for particles that are not spherical in shape. The shape factor "*n*" is introduced to consider the particle geometry.

The thermal conductivity, as determined by the Hamilton-Crosser model, can be expressed as:

$$k_{\text{Eff}} = k_{\text{BF}} \left(\frac{k_{\text{Np}} + (n-1)k_{\text{BF}} - (n-1)\varphi(k_{\text{BF}} - k_{\text{NP}})}{k_{\text{Np}} + (n-1)k_{\text{BF}} + \varphi(k_{\text{BF}} - k_{\text{NP}})} \right) \tag{4.3}$$

where "*n*" is a shape factor:

$n = 3$ for spherical shaped particles
$n > 3$ for elongated particles
$n < 3$ for flattened particles

4.9.2.1 Applications

- *Non-spherical particles*: Pertaining to nanofluids containing nanoparticles that have a shape other than a sphere, such as rods or platelets.
- Higher volume fractions can be accommodated by this model compared to the Maxwell model, because to its shape factor.
- *Industrial applications*: Employed in sectors where the morphology of nanoparticles can be manipulated and enhanced, such as in heat exchangers and cooling systems.

4.9.2.2 Limitations

- *Shape factor estimation*: Accurately determining the shape factor *n* can be challenging.

- Moderate concentrations are not suitable for conditions with very high concentrations, as the interactions between particles become substantial.
- *Empirical nature*: This model necessitates empirical changes and is less theoretical compared to other models such as the Bruggeman model.

4.9.3 BRUGGEMAN MODEL

The Bruggeman model is a sophisticated model that takes into consideration larger volume fractions and the interactions between particles. The model regards the nanofluid as a continuous substance and takes into account the effects of both the base fluid and the nanoparticles.

- The implicit form of this equation usually necessitates the use of numerical methods to solve for k_{eff}.
- Accounts for interactions between nanoparticles and their distribution within the fluid.

The Bruggeman model for a two-phase system can be mathematically represented as:

$$\varphi \left(\frac{k_{\text{NP}} - k_{\text{eff}}}{k_{\text{NP}} + 2k_{\text{eff}}} \right) + (1 - \varphi) \left(\frac{k_{\text{BF}} - k_{\text{eff}}}{k_{\text{BF}} - k_{\text{eff}}} \right) = 0 \tag{4.4}$$

4.9.3.1 Applications
- High volume fractions are ideal for nanoparticle concentrations that are high because they take into account the interactions between particles.
- *Complex systems*: Employed in intricate nanofluid systems where both the base fluid and nanoparticles make substantial contributions to thermal conductivity.
- *Advanced applications*: Suitable for advanced thermal management systems, electronics cooling, and automotive applications where achieving high efficiency is essential.

4.9.3.2 Limitations
- *Complexity*: The inherent nature of the model frequently necessitates the employment of numerical techniques for solving, hence increasing its level of complexity.
- *Parameter sensitivity*: The exact measurement of input parameters greatly affects the performance of this method, making it difficult to implement in some real situations.
- The homogeneous assumption assumes that the mixture is uniform, but this may not be accurate when nanoparticles clump together or form clusters.

4.10 CHALLENGES AND FUTURE DIRECTIONS

Despite significant advancements, challenges remain in the field of nanofluids:

There are two primary limitations that restrict the extensive utilization of nanofluids in large-scale heat exchangers, hence obstructing their commercialization and deployment. Firstly, the presence of solid particles may have detrimental physical impacts on the operation of the prospective working systems. The nanoparticles described in the existing literature have a notable resemblance to the substances responsible for the internal fouling of heat exchanger pipes. The variability in heat transfer enhancement calculations across different literature sources is causing concern over the uncertainty in the results. Since 2001, a cumulative number of 24 patents have been granted for the utilization of nanofluids in electronics and processes. These patents cover a broad spectrum of uses, such as solar collectors and microprocessor cooling applications, each with its own unique set of difficulties. Currently, nanofluid-based products have not had any significant influence on the market.[5]

The market price of nanoparticles plays a crucial role in determining the availability of the product. The increased prices of nanofluids are influenced by various factors, including material charges, preparation costs (such as reagents, surfactants, ultrasonication, and magnetic stirrers), and labor expenses. Hence, the lack of adequate knowledge and data poses a significant challenge in accurately ascertaining the cost of nanofluids, thereby creating an extra barrier. Moreover, the matter of nanoparticle stability in nanofluids is a topic of contention among researchers specializing in nanofluids.[6] The Zeta potential test is employed by researchers to evaluate the nanoparticles' capacity to maintain their dispersion and prevent precipitation. Nevertheless, the endurance and sustained dependability of nanofluids have yet to be substantiated. Furthermore, the long-lasting effects of nanofluids, including their potential harm to both humans and the environment, are not fully understood. Despite the announcement of significant advancements in nanofluids research, there is still a dearth of thorough studies on the prolonged exposure and interactions of nanofluids within specific systems. Moreover, it is essential to evaluate nanofluids in order to determine their capacity to maintain the necessary characteristics under different operational circumstances.[7] To persuade industry and end users to adopt nanofluids as a replacement for conventional heat transfer fluids, it is essential to provide a substantial body of evidence that showcases both the stability and effectiveness of nanofluids.[8] Figure 4.7 provides a thorough overview of the essential procedures and measures needed to bring nanofluids into commercial use, as well as the obstacles that need to be overcome.

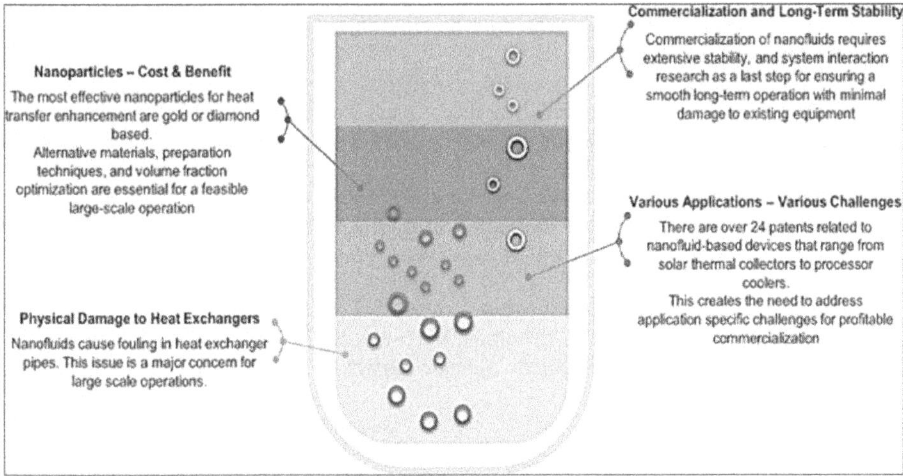

FIGURE 4.7 An important observation on nanofluids for commercialization.[9]

Future research should focus on addressing these challenges to unlock the full potential of nanofluids in various applications.

4.11 SUMMARY

- Nanofluids represent a significant advancement in the field of thermal management and fluid dynamics, offering enhanced properties that surpass traditional fluids.
- The fundamentals of nanofluids, the latest methods for their preparation and characterization, and the factors influencing their thermophysical properties have been discussed.
- Theoretical models and recent research trends provide deeper insights into their behavior and potential applications.
- Overcoming the challenges associated with stability, scale-up, and environmental impact will be crucial for the widespread adoption of nanofluids in industrial and engineering applications.

REFERENCES

1. Choi, S. U., & Eastman, J. A. (1995). *Enhancing Thermal Conductivity of Fluids with Nanoparticles* (No. *ANL/MSD/C*P-84938; CONF-951135-29). Argonne National Lab: Argonne, IL.
2. Mbambo, M. C., Khamlich, S., Khamliche, T. et al. (2020). Remarkable thermal conductivity enhancement in Ag—decorated graphene nanocomposites based nanofluid by laser liquid solid interaction in ethylene glycol. *Scientific Reports*, *10*(1), 10982. doi:10.1038/s41598-020-67418-3
3. Munyalo, J. M., & Zhang, X. (2018). Particle size effect on thermophysical properties of nanofluid and nanofluid based phase change materials: A review. *Journal of Molecular Liquids*, *265*, 77–87.
4. Ali, H. M., Babar, H., Shah, T. R., Sajid, M. U., Qasim, M. A., & Javed, S. (2018). Preparation techniques of TiO2 nanofluids and challenges: a review. *Applied Sciences*, *8*(4), 587.
5. Malika, M., Pargaonkar, A., & Sonawane, S. S. (2023). Experimental and statistical analysis of MWCNT hybrid nanofluid-based multi-functional drilling fluid. *Chemical Papers*, *77*(11), 6773–6784.
6. Malika, M., & Sonawane, S. S. (2024). Ecological optimization and LCA of TiO2-SiC/water hybrid nanofluid in a shell and tube heat exchanger by ANN. *Proceedings of the Institution of Mechanical Engineers, Part E: Journal of Process Mechanical Engineering*, *238*(1), 45–55.
7. Malika, M., Pargaonkar, A., & Sonawane, S. S. (2024). Performance of an emulsion nanofluid membrane for the extraction of antimony heavy metal: experimental and numerical investigation. *Journal of Sustainable Metallurgy*, 10, 460–471.
8. Malika, M., & Sonawane, S. (2024). A Review on the application of nanofluids in enhanced oil recovery. *Current Nanoscience*, *20*(3), 328–338.
9. Alami, A. H., Ramadan, M., Tawalbeh, M. et al. (2023). A critical insight on nanofluids for heat transfer enhancement. *Science Report*, *13*, 15303. https://doi.org/10.1038/s41598-023-42489-0

5 Ultrasonic-Aided Fabrication of Nanofluids

5.1 PRINCIPLES OF ULTRASONIC-AIDED FABRICATION: KEY MECHANISMS

Ultrasonic processing utilizes ultrasound, usually within the frequency range of 20 kHz to several MHz. The process is dependent on the occurrence of acoustic cavitation, which refers to the creation, expansion, and sudden collapse of bubbles within a liquid. This structural failure results in highly concentrated heat and pressures, resulting in powerful shock waves and micro-jets. The aforementioned impacts have the capability to disintegrate clusters of nanoparticles and disperse them evenly inside the primary fluid.

5.2 CAVITATION

Cavitation is essential in the synthesis and operation of nanofluids, which are mixtures of nanoparticles and a base fluid. Cavitation is the process in three steps, which includes:

- Bubbles formation and expansion,
- Collapse of the bubbles, and
- Formation of shock waves and micro-jets.

As dispersing nanoparticles uniformly within the base fluid is crucial for increasing the physical and thermal properties of nanofluids, gaining a comprehensive understanding of the significance of cavitation is crucial for optimizing the fabrication process and enhancing the stability and performance of nanofluids in many applications.

5.3 MECHANISMS OF CAVITATION IN NANOFLUID FABRICATION

Cavitation is a complex phenomenon that encompasses multiple stages, each of which plays a vital role in the process of using ultrasonic waves to create nanofluids. The main processes involved are the *formation and enlargement of bubbles*, the *implosion of bubbles*, and the resulting production of *shock waves and micro-jets*. These mechanisms together lead to the efficient scattering and stability of nanoparticles within the main fluid.

DOI: 10.1201/9781003594949-5

5.3.1 Formation and Expansion of Bubbles

Cavitation is initiated by the formation of bubbles in a liquid, caused by the transmission of ultrasonic waves through the substance. The procedure entails the subsequent stages.

5.3.1.1 Stage 1: Nucleation Sites

Zettlemoyer[1] (1969) and Burton[2] (1977) define liquid-gas nucleation as the spontaneous creation of bubbles within a liquid. This process, known as subcooling, happens when temperature and pressure are kept constant and are lower than the saturation pressure of a pure liquid. Bubbles are created at specific locations called *nucleation sites*. These sites can be tiny gas pockets, impurities, or small openings in solid surfaces present in the liquid. Nucleation can occur in two forms: homogeneous or heterogeneous.

Homogeneous nucleation refers to the formation of cavitation bubbles in a uniform manner. Where, cavitation bubbles form uniformly throughout the liquid, without any specific locations or contaminants. This procedure necessitates a substantial amount of tensile tension (negative pressure) or intense superheating, which makes it less prevalent in practical situations due to the formidable energy barrier it entails.

Key points

- The energy barrier is exceptionally high due to the necessity of creating a bubble within a homogeneous and uncontaminated liquid.
- *Requirements*: Demands either high negative pressure or superheating in order to surpass the energy barrier.
- *Frequency*: Infrequent as a result of the rigorous requirements involved.
- In a highly purified liquid, the formation of bubbles throughout the liquid due to a large decrease in pressure below the vapor pressure is an uncommon occurrence in most practical scenarios.

In the context of cavitation, which is the formation and collapse of vapor bubbles in a liquid, *heterogeneous nucleation* occurs when vapor bubbles form preferentially at surfaces or interfaces, such as on impurities, suspended particles, or the walls of a container, where the energy barrier for bubble formation is lower compared to the bulk liquid. The occurrence of heterogeneous nucleation of bubbles is more prevalent and takes place at specific locations such as surfaces, contaminants, or tiny gas pockets inside the liquid. These locations reduce the amount of energy needed for the creation of bubbles, thus facilitating the process of bubble nucleation.

Key points

- The energy barrier for nucleation is reduced when there are surfaces or impurities present, compared to the energy barrier for homogeneous nucleation.
- *Conditions*: Takes place under milder circumstances compared to homogenous nucleation.
- *Frequency*: More prevalent in real-life situations.

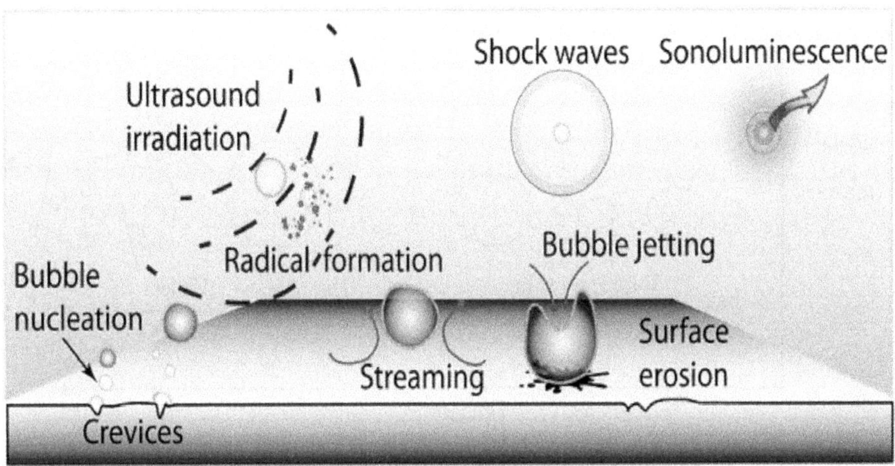

FIGURE 5.1 Cavitation principle: Bubble nucleation and dynamics.[3]

- In a pump or turbine, cavitation bubbles frequently develop at surface defects, contaminants, or pre-existing microbubbles that are trapped on surfaces. These spots facilitate the development of bubbles by offering a more efficient conduit for energy transfer (Figure 5.1).

5.3.1.2 Stage 2: Alternating Pressure Cycles

Ultrasound waves are composed of alternating cycles of high pressure (compression) and low pressure (rarefaction). During the low-pressure cycles, the pressure of the liquid decreases below its vapor pressure, resulting in the formation of vapor cavities or bubbles.

Cavitation is the phenomenon in which rapid fluctuations in pressure within a liquid cause the creation, enlargement, and subsequent implosion of tiny gas bubbles. The phenomenon is greatly affected by the alternating cycles of high-pressure (compression) and low-pressure (rarefaction) ultrasonic waves.

- *Compression phase (high pressure)*: The compression phase (high pressure) involves the compression of liquid molecules by ultrasonic waves, resulting in an elevation of local pressure.
- *Rarefaction phase (low pressure)*: The presence of high pressure inhibits the development of gas bubbles.

In the rarefaction phase, the ultrasonic waves induce areas of reduced pressure within the liquid. Gas bubbles are formed when the pressure falls below the vapor pressure of the liquid.

5.3.1.3 Stage 3: Bubble Growth

After nucleation, bubbles expand by absorbing additional vapor or gas from the surrounding liquid during subsequent low-pressure cycles. The bubble development

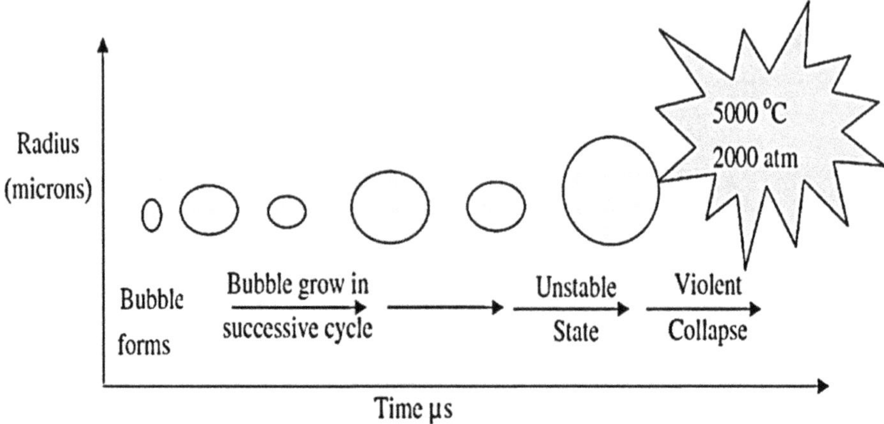

FIGURE 5.2 Schematic representation of a cavitation bubble formation and its growth stage.[4]

stage is a crucial component of the cavitation process, in which the initial small gas bubbles generated during the rarefaction phase enlarge in size as a result of the ongoing impact of the alternating pressure cycles of the ultrasonic waves. The growth can persist for multiple sonic cycles, contingent upon the intensity and frequency of the ultrasound. A visual representation to illustrate the bubble growth stage in cavitation is shown in Figure 5.2. This stage can be divided into multiple sub-processes that take place as a result of low-pressure cycles.[5]

i. **Initial Bubble Formation**
   - During the rarefaction phase, the pressure in the liquid decreases to a level below its vapor pressure, resulting in the release of dissolved gases from the solution and the formation of small gas bubble nuclei.
   - The early bubbles are commonly little, usually measuring in the micron or sub-micron scale.

ii. **Primary Growth Mechanism**
   - As the ultrasonic vibrations persist, the liquid experiences several cycles of rarefaction and compression.
   - During each cycle of rarefaction, the pressure surrounding the bubbles decreases, thereby diminishing the cohesive forces between the molecules of the liquid.
   - The decrease in pressure enables the bubbles to enlarge, since the gas inside the bubble can exert a stronger outward push than the pressure of the liquid can confine.

iii. **Boyle's Law and Gas Diffusion**

$$\text{Boyle's Law}: P_1V_1 = P_2V_2 \tag{5.1}$$

   - Boyle's Law defines the correlation between the pressure (P) and volume (V) of a gas when the temperature remains constant.

- During the rarefaction process, the gas bubbles experience an increase in volume as the external pressure lowers, in accordance with Boyle's Law.
- Furthermore, gas diffusion also contributes to the formation of the bubble by allowing dissolved gases in the liquid to seep into it.

iv. **Rectified Diffusion**
- Rectified diffusion is a phenomenon in which gas molecules exhibit more efficiency in diffusing into a bubble during the rarefaction phase compared to their departure during the compression phase.
- This imbalance results in a cumulative rise in the gas content of the bubble across subsequent cycles, resulting in continual growth.

v. **Thermal Effects**
- The formation of bubbles can be influenced by temperature fluctuations resulting from oscillating pressure.
- During the process of rarefaction, the temperature within the bubble may decrease somewhat, causing a greater amount of gas to dissolve into the bubble.
- On the other hand, when compressing, the temperature increases, but this impact is usually uneven, resulting in a greater overall expansion during rarefaction.

vi. **Dynamic Stability**
- The stability of the expanding bubble is upheld by the equilibrium between the forces exerted on it: the internal liquid pressure and surface tension counterbalanced by the external gas pressure.
- If the bubble expands rapidly and reaches a significant size, it can become unstable and perhaps break apart.

Key Factors that Affect the Formation and Growth of Bubbles

- *Ultrasonic frequency*: Lower frequencies, usually about 20 kHz, are more efficient in creating larger bubbles because they have longer wavelengths and more noticeable pressure changes.
- *Ultrasonic power*: Increasing the power levels results in the production of more intense acoustic waves, which in turn leads to the formation of larger bubbles and faster growth.
- *Temperature*: Elevated liquid temperatures reduce surface tension and enhance vapor pressure, making bubble formation and expansion simpler.
- *Base fluid properties*: The viscosity and surface tension of the base fluid have an effect on the behavior of bubbles. Fluids with lower viscosity typically promote more intense bubble action.

5.3.2 COLLAPSE OF BUBBLES

Key events in collapse include:

- *Bubble contraction*: The bubble undergoes a rapid reduction in size when the pressure from the outside becomes greater than the pressure of the vapor inside.

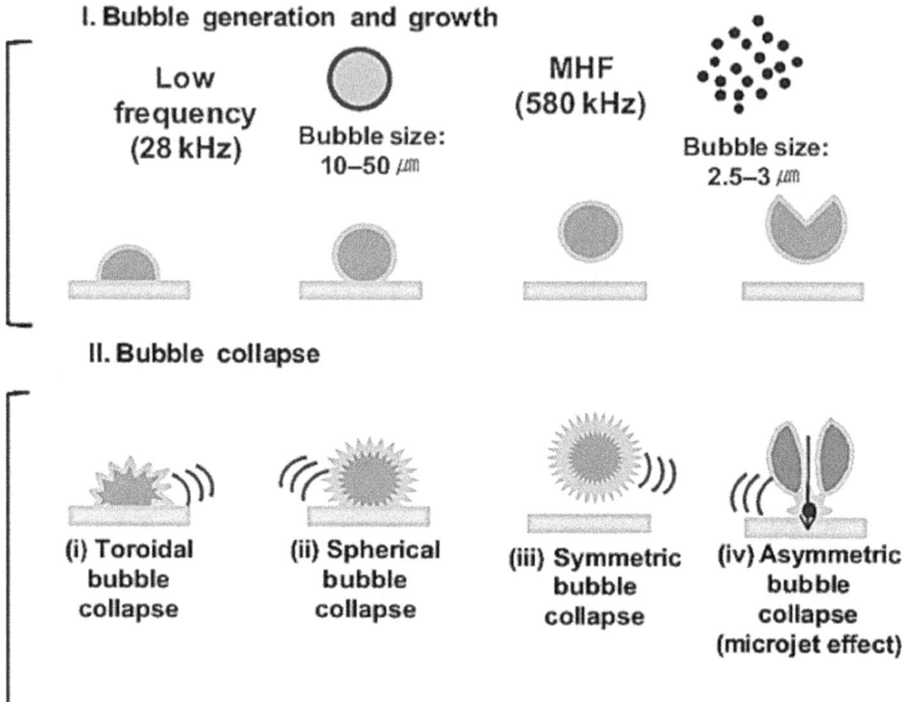

FIGURE 5.3   Symmetric and asymmetric collapse of cavitation bubbles.[6]

- *Shock wave generation*: The collapse of the bubble produces shock waves that travel through the liquid.
- *Micro-jet formation*: Under conditions of asymmetry, the collapse can generate high-velocity micro-jets that are aimed at adjacent surfaces.
- *Localized high temperature and pressure:* The collapse results in the generation of exceedingly elevated temperatures and pressures at the implosion site, which enables the disintegration of materials into nanoparticles.

The collapse of cavitation can happen either symmetrically or asymmetrically (Figure 5.3), and it has a substantial impact on the synthesis of nanofluids. Now, let us explore both forms of collapse and their implications.

i. **Symmetric Collapse:**
    Symmetric collapse refers to the uniform collapse of a bubble from all sides. This phenomenon typically occurs when the bubble is located at a considerable distance from any solid barriers, leading to a spherical collapse. The homogeneous collapse generates highly concentrated pressure and temperature, resulting in the creation of nanoparticles from suspended elements in the fluid.

Key Characteristics
- Uniform collapse refers to the sudden and complete failure of a uniform or clothing item, resulting in its loss of shape, structure, or integrity.
- Elevated localized pressure and temperature
- Generation of shock waves

ii. **Asymmetric Collapse:**

Asymmetric collapse refers to the situation where a bubble is in close proximity to a barrier, such as a solid surface. The existence of the boundary results in the bubble collapsing in an uneven manner, frequently creating a high-speed micro-jet that is aimed toward the surface. The micro-jet possesses considerable kinetic energy, resulting in a substantial impact on the surface. This impact causes an intensified disintegration and blending of materials.

Key Characteristics
- Asymmetrical collapse
- Micro-jets are created through the process of formation.
- Improved surface collision and blending

Key Factors that Affect the Formation and Growth of Bubbles

The growth and violent collapse of cavitation bubbles in an acoustic cavitation process are greatly affected by the characteristics of ultrasound, such as the type of transducer (either a bath or probe type sonicator), the power, intensity, and frequency and temperature and base fluid properties. It is widely acknowledged that the frequency of ultrasound has a substantial impact on the occurrence of cavitation. With the exception of frequencies over a few megahertz (MHz), a cavitation bubble theoretically expands during the negative phase of the acoustic cycle and is compelled to collapse during the positive phase. At low frequencies (~20 kHz), cavitation bubbles experience growth for about 25 µs when exposed to equal acoustic intensities and pressure amplitudes.

However, at high acoustic frequencies (~1 MHz), the growth of cavitation bubbles is limited to only 0.5 µs. The variations in bubble growth lead to smaller bubble sizes at higher frequencies, resulting in less intense collapses. Specifically, at low frequencies like 28 kHz, the bubble size is roughly 10–50 µm, whereas at medium-high frequencies like 580 kHz, the bubble size is around 2.5–3.0 µm, as depicted in Figure 5.3.

Impact of Bubble Collapse

- *Localized hot spots*: The intense temperatures and pressures generate specific areas of high temperature known as "hot spots", which have the ability to trigger chemical reactions or alter the surface chemistry of nanoparticles.
- High shear forces are produced by the implosive collapse, which is crucial for disintegrating nanoparticle agglomerates in the surrounding liquid.

Nanofluid Synthesis Mechanism

- *Energy transfer*: The energy that is released when a bubble collapses is transmitted to the particles that are suspended in the fluid.

- Particle breakage occurs when the combination of high pressure and temperature causes bigger particles to fracture, resulting in the formation of smaller nanoparticles.
- *Improved mixing*: The impact waves and micro-jets enhance the process of mixing, resulting in a consistent dispersion of nanoparticles.

5.3.3 Shock Waves and Micro-Jets

The explosion of these bubbles generates secondary phenomena such as shock waves and micro-jets, which are essential in promoting the dispersion of nanoparticles in the fluid. Shock waves are crucial in the process of using ultrasound to create nanofluids. Shock waves play a crucial role in dispersing nanoparticles, which in turn improves the characteristics and stability of nanofluids. These shock waves are generated, have an influence, and interact with each other, contributing to the overall effectiveness of dispersion.

5.3.3.1 Generation of Shock Waves

Shock waves are produced as a result of the sudden and forceful collapse of cavitation bubbles in a liquid medium exposed to ultrasonic vibrations.

- Ultrasonic waves induce the formation of small vapor bubbles in a liquid by subjecting it to alternating cycles of high-pressure compression and low-pressure rarefaction. These bubbles emerge during the rarefaction phase when the local pressure decreases.
- These bubbles increase in size during successive cycles through a process called rectified diffusion, wherein gas diffuses into the bubble more during the low-pressure phase than it diffuses out during the high-pressure phase.
- As the compression phase occurs, the liquid pressure in the surrounding area experiences a rapid increase, leading to a catastrophic collapse of the bubbles. This collapse is very accelerated, frequently happening within microseconds.
- Formation of shock waves: The collapse of the bubble causes the gas inside it to be compressed to exceedingly high pressures. The quick compression results in the formation of a shock wave, which is a powerful acoustic wave that spreads outwards from the location of the collapse.
- Repeated bubble collapses generate numerous shock waves, creating a highly dynamic environment that continually agitates and mixes the nanoparticles.

5.3.3.1.1 *Isotropic Bubble Collapse and Shock Wave Propagation*

Isotropic bubble collapse in the context of acoustic cavitation refers to the symmetrical implosion of a bubble, leading to the production of shock waves that propagate uniformly in all directions. This effect is essential for the scattering of nanoparticles in the process of using ultrasound to create nanofluids. In this study, we investigate the intricate processes involved in the collapse of isotropic bubbles and the subsequent propagation of shock waves.

5.3.3.1.2 *The Mechanism of Isotropic Bubble Collapse*

- Vapor bubbles are created and expand in a uniform liquid environment when exposed to ultrasonic waves, due to the changing pressure cycles.
- The shape of these bubbles is usually spherical because of the isotropic characteristics of the pressure field.
- During the rarefaction phase, which is the cycle of low pressure, the bubbles increase in size when the local pressure decreases below the vapor pressure of the liquid.
- The bubble expands by gathering vapor or gas from the adjacent liquid.
- During the compression phase of the high-pressure cycle, the bubbles initiate a process of collapse as the pressure increases.
- During an isotropic collapse, the bubble undergoes a symmetrical implosion, with all directions converging toward its center.
- The collapse's symmetrical nature arises from the evenly distributed pressure field that surrounds the bubble.
- The implosion of the bubble causes the energy to be focused into a confined space, leading to exceedingly elevated temperatures and pressures in the center of the bubble.
- The concentration of energy is crucial for the generation of shock waves as shown in Figure 5.4a.
- The energy emanating from the isotropic collapse is evenly dispersed throughout the bubble, resulting in the formation of symmetrical shock waves.

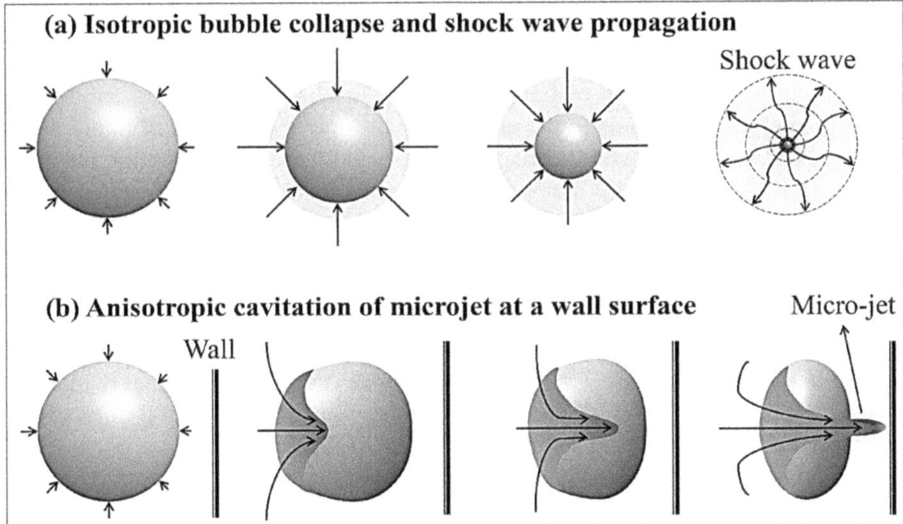

FIGURE 5.4 Illustration of (a) isotropic bubble collapse and shockwave propagation; (b) anisotropic cavitation of micro-jet.[6]

- Localized heating occurs when the collapse generates specific areas of elevated temperature. This increase in temperature can have an impact on chemical reactions and the characteristics of particle surfaces.

5.3.3.1.3　Propagation of Shock Waves

- The swift disintegration of the bubble results in an abrupt surge in pressure at the location of the disintegration.
- This sudden increase in pressure generates a shock wave, which is a powerful acoustic wave that travels through the liquid.
- Because the collapse is isotropic, the shock wave spreads uniformly in all directions from the collapse point.
- The shock wave propagates radially as a spherical wave, transmitting energy through the liquid.

5.3.3.1.4　Impact of Nanoparticle Dispersion

- *Cluster penetration*: Shock waves with high pressure can infiltrate clusters of nanoparticles, causing the disruption of intermolecular interactions and the dispersion of the particles.
- *Improved mixing*: The symmetrical spread of shock waves guarantees the even distribution of nanoparticles in the liquid, eliminating clumping and settling.
- *Surface modification*: The extreme circumstances produced by shock waves can modify the surface characteristics of nanoparticles, improving their interaction with the underlying fluid.

5.3.3.2　Anisotropic Cavitation and the Formation of High-Speed Micro-Jets

Anisotropic cavitation, also known as un-isotropic cavitation, describes the nonuniform (asymmetric) collapse of bubbles, usually in close proximity to solid limits or other bubbles. This leads to the creation of fast-moving streams of liquid called micro-jets. Anisotropic bubble collapse in the field of acoustic cavitation results in the creation of high-velocity micro-jets. The micro-jets possess sufficient power to greatly influence the dispersion of nanoparticles in nanofluids. In this discussion, we will provide detailed information regarding the creation, speed and strength, and interaction with surfaces of these little jets.

5.3.3.2.1　Formation of Micro-Jets

- Micro-jets are created when cavitation bubbles burst in an unequal manner. The presence of solid surfaces or other bubbles in the vicinity of the collapsing bubble typically causes this asymmetry (Figure 5.5).
- The formation process involves the creation of cavitation bubbles within a liquid when it is exposed to ultrasonic waves, resulting from the alternating cycles of high and low pressure.
- These bubbles undergo initial expansion during the low-pressure cycles.
- When a bubble is in close proximity to a solid surface or another bubble, its collapse becomes asymmetric as a result of the unequal distribution of pressure surrounding the bubble.

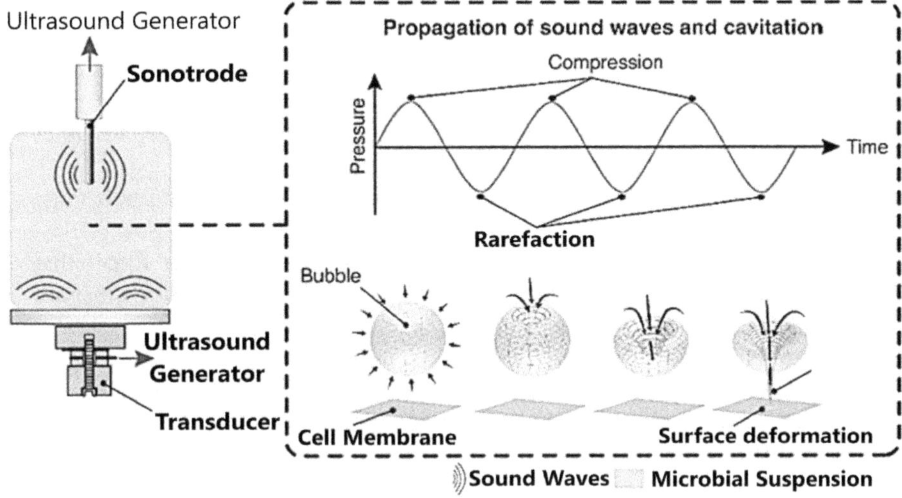

FIGURE 5.5 Process of micro-jet formation resulting from the implosion of acoustic cavitation bubbles.[7]

- The side of the bubble that is in contact with a solid surface or another bubble experiences a slower collapse rate compared to the opposite side.
- The bubble experiences a faster collapse on the side opposite to the surface, causing the liquid on that side to accelerate toward the side that collapses more slowly.
- This leads to the creation of a rapid liquid stream (micro-jet) that propels outward from the bubble toward the surface.

5.3.3.2.2 Characteristics of Micro-Jets

- Micro-jets may reach velocities of several hundred meters per second, reaching up to approximately 200 m/s or even higher.
- The high velocity is a result of the swift and concentrated motion of the liquid as it moves from one side of the collapsing bubble to the other.
- Upon collision, the high-speed jet imparts a focused and intense amount of kinetic energy to a localized region.
- The micro-jets generate a significant force that might result in mechanical damage or alteration to surfaces and particles they come into contact with.

5.3.3.2.3 Quantifying the Impact

- Velocity: v is about between 100 and 200 m/s.
- The magnitude of the force is contingent upon the velocity of the jet, the density of the liquid, and the surface area of contact. A typical micro-jet can generate localized pressures of several megapascals (MPa).

5.3.3.2.4 Surface Interaction of Micro-Jets

Micro-jets have a strong interaction with solid surfaces and clusters of nanoparticles, resulting in several effects that help disperse and stabilize nanoparticles in nanofluids.

Surface Interaction Effects:

- When micro-jets collide with solid surfaces, the forceful impact at high speeds can lead to concentrated erosion in specific areas.
- Erosion has the ability to alter the physical appearance of a surface by either making it rougher or by removing material from it.
- Micro-jets can generate strong shear stresses to destroy nanoparticle clusters (agglomerates).
- The kinetic energy generated by the jet disrupts the cohesive forces that bind the agglomerates, leading to the formation of smaller particles that are evenly distributed.
- Micro-jets cause turbulence and create small areas of mixing.
- The improved blending facilitates the even dispersion of nanoparticles within the base fluid, resulting in greater stability and uniformity of the nanofluid.

5.3.3.2.5   Applications of Shock Waves and Micro-Jets
- *Nanoparticle dispersion*: Shock waves and micro-jets produce powerful mechanical forces that efficiently disperse nanoparticles evenly throughout the base fluid.
- Surface modification refers to the process of changing the surface chemistry of nanoparticles. This is done by subjecting them to extreme conditions, which might enhance their stability and their capacity to interact with the base fluid.
- *Improvement of physical properties*: The uniform distribution of nanoparticles in nanofluids greatly improves their thermal, rheological, and mechanical properties.

5.4   KEY BENEFITS OF CAVITATION IN NANOFLUIDS PROPERTIES

5.4.1   Deagglomeration of Nanoparticles

- Agglomeration occurs when nanoparticles, because of its dimensions, possess a substantial surface area to volume ratio, resulting in notable van der Waals forces and electrostatic attractions among them. This results in the particles coalescing and producing agglomerates.
- Cavitation is the occurrence of small vapor-filled cavities or bubbles in a liquid due to rapid variations in pressure. When these bubbles implode, they produce intense shear forces.
- The powerful shear forces generated by the bursting cavitation bubbles are sufficiently robust to surpass the attraction forces that bind the nanoparticles together. As a consequence, the nanoparticle clusters are disintegrated (deagglomerated), resulting in a more uniform distribution throughout the base fluid as shown in Figure 5.6.
- Direct ultrasonication (using a horn/probe) is typically more efficient in dispersing dry nanoparticle powder into a base fluid. This is because it directly

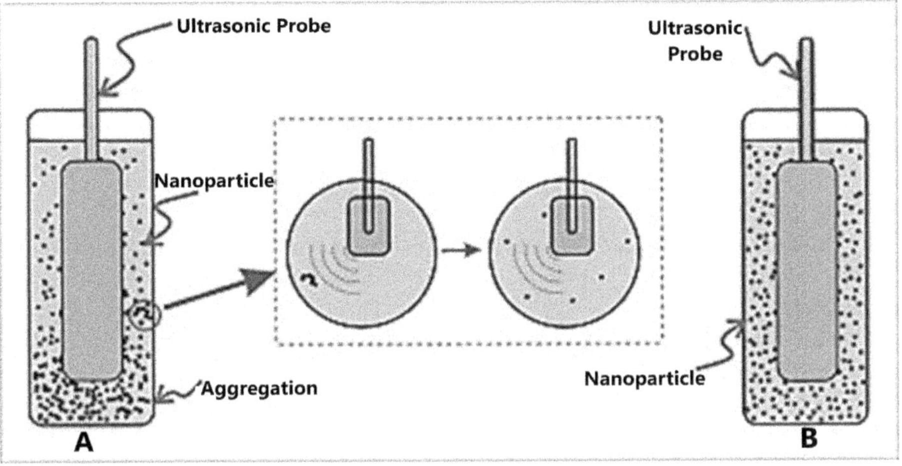

FIGURE 5.6 Cavitation for the deagglomeration of nanoparticles in base fluid.[8]

applies intense ultrasonic energy to the fluid, resulting in the effective separation of particle clusters and complete mixing. Enhanced effectiveness is achieved by the exact regulation of sonication settings.

- On the other hand, indirect ultrasonication (ultrasonic bath) has a lower intensity and less direct energy transfer, which makes it less efficient in breaking up clusters of nanoparticles and achieving a consistent dispersion.

5.4.2 EFFECTIVE ULTRASONICATION PROCESS FOR BETTER COLLOIDAL DISPERSION OF NANOFLUID

Ultrasonication is a commonly employed method for evenly distributing nanoparticles in fluids to form stable colloidal dispersions, such as nanofluids. The efficacy of ultrasonication in attaining a consistent and enduring dispersion relies on various crucial factors, namely ultrasound frequency, amplitude, and ultrasonication time. Below is a comprehensive analysis of how each parameter impacts the process:

5.4.2.1 Frequency of Ultrasound

Frequency is the measurement of the number of cycles of sound waves per second, expressed in kilohertz (kHz). The typical frequency range for ultrasonication in colloidal dispersion is between 20 kHz and several hundred kHz.

5.4.2.1.1 Low-Frequency Ultrasound (20–40 kHz)

Ultrasound with a low-frequency range of 20–40 kHz generates high-intensity cavitation bubbles that collapse forcefully, resulting in the production of powerful shear forces. These forces have the ability to disintegrate clusters and distribute nanoparticles efficiently. Nevertheless, an excessive level of intensity might occasionally result in particle degradation or undesired chemical reactions.

5.4.2.1.2 *High-Frequency Ultrasound (>100 kHz)*

Ultrasound with a frequency greater than 100 kHz generates smaller cavitation bubbles that collapse with less intensity. This can lead to a gentler blending process and is frequently employed for delicate substances, where strong shear pressures could potentially harm the particles or the underlying fluid.

5.4.2.2 Amplitude

Amplitude is the highest level of vibration or the amount of energy produced by the ultrasonic probe, which is generally linked to the power setting on the ultrasonicator.

Higher amplitude: A greater amplitude leads to more intense cavitation and stronger shear forces, which can fragment bigger aggregates and improve dispersion. Nevertheless, a significant rise in amplitude can also heighten the likelihood of overheating the sample, which could potentially result in the deterioration or alteration of both the nanoparticles and the base fluid's characteristics.

Lower amplitude: Decreasing the amplitude results in a more mild mixing action, which is advantageous for handling fragile substances. However, it may necessitate extended processing durations to obtain a consistent dispersion.

Optimal Amplitude

- Discovering the ideal amplitude is essential for attaining a harmonious equilibrium between efficient dispersion and mitigating the risk of overheating or harm. This typically entails commencing at a moderate level of intensity and making gradual modifications based on the observed particle size distribution and stability of the nanofluid.
- *Monitoring and temperature regulation*: Introducing a cooling mechanism while using ultrasonication can effectively control the heat produced at large amplitudes, enabling efficient dispersion without causing thermal harm.
- Characterization refers to the process of creating and developing characters in a story, play, or other literary work. Methods such as dynamic light scattering (DLS), transmission electron microscopy (TEM), or scanning electron microscopy (SEM) are frequently employed to evaluate the distribution of particle sizes and verify the effectiveness of the dispersion process.

Sample Study

In this study, alumina nanoparticles are evenly distributed in water by utilizing ultrasonication (Figure 5.7).

- *Small amplitude (20%)*: The mean particle size may stay relatively big (e.g., 200 nm) with a wider range of sizes.
- With a moderate amplitude of 50%, the average particle size is likely to fall significantly, for example, to 100 nm, resulting in a smaller size distribution. This suggests improved dispersion.

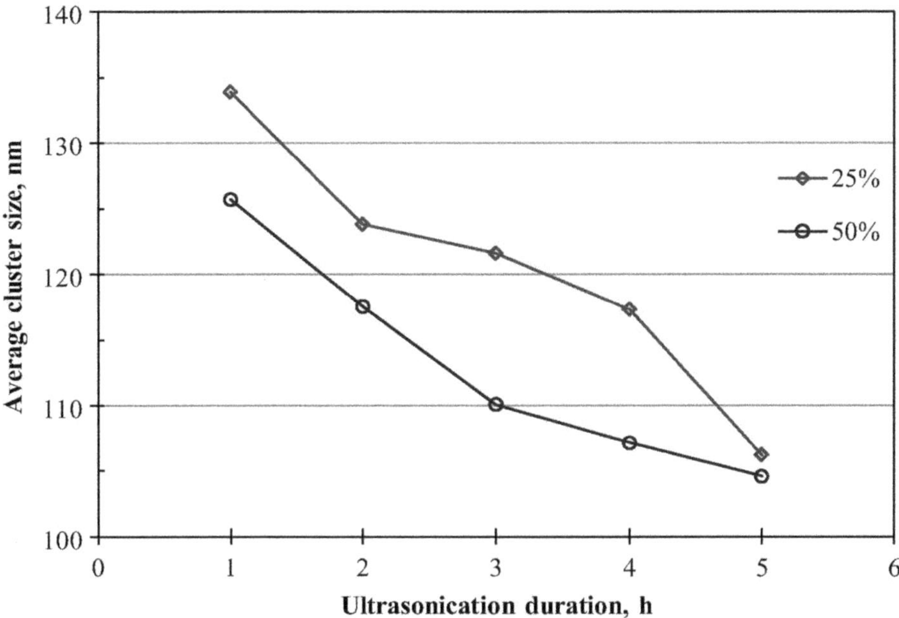

FIGURE 5.7 Alumina–water nanofluid's average particle size after 25% and 50% ultrasonication amplitudes.[9]

- With a high amplitude of 80%, there is a possibility of further reducing the average particle size, perhaps to 50 nm. However, it is crucial to carefully control the heat generated to prevent any detrimental impact on the quality or stability of the dispersion.

5.4.2.3 Ultrasonication Time

- Ultrasonication is the process of using high-frequency sound waves to create mechanical vibrations in a liquid or solid material. Time refers to the period during which the sample is subjected to ultrasonication.
- Inadequate ultrasonication duration may not supply sufficient energy to disintegrate all aggregates, resulting in inadequate dispersion.
- Extended periods of ultrasonication can result in improved dispersion but also carry the potential risks of overheating and probable deterioration of the nanoparticles and fluid. If not adequately regulated, extended ultrasonication can occasionally lead to re-agglomeration as a result of particle-particle collisions.

5.4.2.4 Optimal Parameters

- The ideal ultrasonication settings are determined by the unique attributes of the nanoparticles and the base fluid, along with the intended qualities of the resulting nanofluid. Here is a comprehensive framework for optimizing the process:

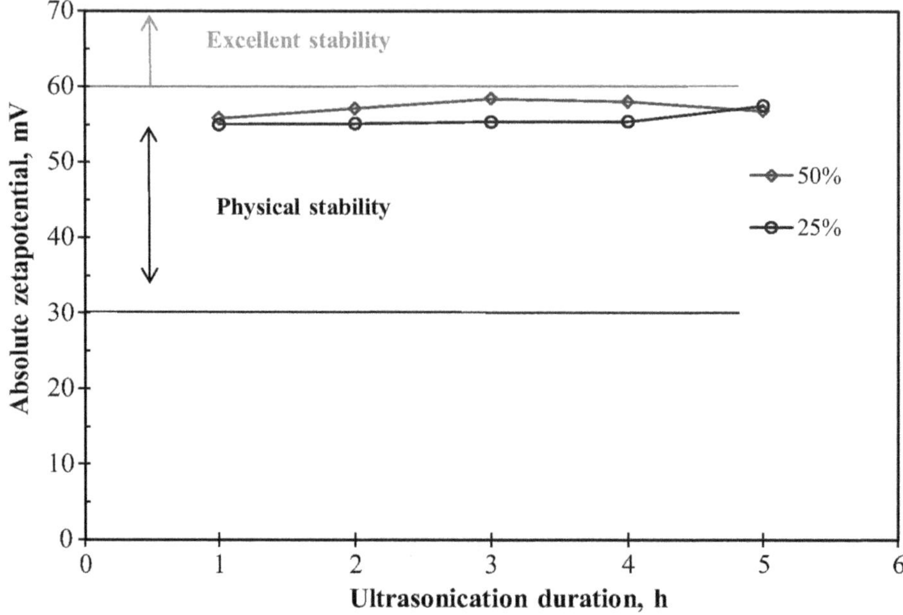

FIGURE 5.8 Alumina–water nanofluid absolute zeta potentials after 25% and 50% ultrasonication amplitudes.[9]

- *Commence with a moderate frequency and amplitude*: Initiate with a frequency ranging from 20 to 40 kHz and an amplitude set at a level in the middle range. Modify according to initial findings.[10]
- Temperature monitoring is necessary because ultrasonication can produce substantial heat. Employ cooling techniques as needed to avoid excessive warmth.
- Gradually increase the duration of ultrasonication, starting with a few minutes, and closely observe the quality of dispersion. Conduct regular inspections to prevent excessive processing.
- *Assess the level of dispersion quality*: Employ methodologies such as DLS, zeta potential testing (Figure 5.8), and visual examination (e.g., TEM or SEM imaging) to evaluate the dispersion condition and stability.

5.5 PRACTICAL EXAMPLE

- *Nanoparticles*: Titanium dioxide (TiO_2)
- The *base fluid* options include water or ethylene glycol.
- The *initial parameters* for ultrasonication are as follows: a frequency of 20 kHz, an amplitude of 50%, and a duration of 10 min.
- Optimization involves systematically modifying the frequency and amplitude, while simultaneously monitoring the particle size distribution and stability, until the ideal dispersion is attained.

- By precisely adjusting these parameters, you may attain a consistent and uniform distribution of nanoparticles in the main fluid, hence improving the efficiency and range of uses for the nanofluid.[11]

5.6 SUMMARY

5.6.1 EFFECTIVE ULTRASONICATION PROCESS FOR BETTER COLLOIDAL DISPERSION OF NANOFLUID FOR ENHANCED MIXING AND HOMOGENEITY

- *Turbulent conditions*: Cavitation induces turbulence in the fluid, creating chaotic and vigorous fluid motion.
- *Thorough mixing*: The turbulent conditions ensure that nanoparticles are not just dispersed but are thoroughly mixed throughout the base fluid.
- *Homogeneous distribution*: This process helps achieve a uniform distribution of nanoparticles, which is essential for ensuring that the nanofluid has consistent thermal and physical properties throughout its volume. Uniformity is critical because localized variations in nanoparticle concentration can lead to inconsistent behavior and performance of the nanofluid.

5.6.2 IMPROVED STABILITY

- *Stable dispersions*: Stability refers to the ability of nanoparticles to remain evenly dispersed without settling or separating over time.
- *Intense mixing*: The vigorous mixing from cavitation helps in keeping the nanoparticles suspended and prevents them from settling.
- *Surface modification effects*: The surface modifications induced by cavitation can enhance the interaction between nanoparticles and the base fluid, further promoting stability.
- *Long-term stability*: Stable nanofluids are less prone to issues like sedimentation (settling of particles) and separation (phase separation), which are common problems in nanofluids. Stable dispersions are crucial for the practical applications of nanofluids, ensuring consistent performance over extended periods.

REFERENCES

1. Zettlemoyer, A. C. (1969). *Nucleation*. New York: Marcel Dekker.
2. Burton, J. J. (1977). Nucleation theory. In Bruce J. Berne (Ed.), *Statistical Mechanics: Part A: Equilibrium Techniques* (pp. 195–234). Boston, MA: Springer.
3. Rashmi, P., Swaraj, R. F. C.-Q., & Juan Carlos Colmenares, Q. (2019). Designing microflowreactors for photocatalysis using sonochemistry: a systematic review article. *Molecules, 24*(18), 3315. https://doi.org/10.3390/molecules24183315
4. Madhavan, J., Jayaraman, T., Dhandapani, B., Salla, S., Myong Yong, C., & Muthupandian, A. (2019). Hybrid advanced oxidation processes involving ultrasound: an overview. *Molecules, 24*(18), 3341. https://doi.org/10.3390/molecules24183341

5. Park, S., Park, J. S., Lee, H., Heo, J., Yoon, Y., Choi, K., & Her, N. (2011). Ultrasonic degradation of endocrine disrupting compounds in seawater and brackish water. *Environmental Engineering Research, 16*(3), 137–148.

6. Ren, X., Zheng, T., Yanshan, D., G. Ma, Zhongze, L., Xiangning, B., Muhammad, B., Ali Behrad, V., & Ahmad, H. (2023). Effects of mechanical stirring and ultrasound treatment on the separation of graphite electrode materials from copper foils of spent LIBs: A comparative study. *Separations, 10*(4), 246. https://doi.org/10.3390/separations10040246

7. Peng, K., Mohamed, K., Olivier, B., & Eugène, V. (2020). Recent insights in the impact of emerging technologies on lactic acid bacteria: A review. *Food Research International, 137*, 109544.

8. Asadi, A., Pourfattah, F., Szilágyi, I. M., Afrand, M., Żyła, G., Ahn, H. S.,... & Mahian, O. (2019). Effect of sonication characteristics on stability, thermophysical properties, and heat transfer of nanofluids: A comprehensive review. *Ultrasonics Sonochemistry, 58*, 104701. https://doi.org/10.1016/j.ultsonch.2019.104701

9. Mahbubul, I. M., Saidur, R., Amalina, M. A., Elcioglu, E. B., & Okutucu-Ozyurt, T. (2015). Effective ultrasonication process for better colloidal dispersion of nanofluid. *Ultrasonics Sonochemistry, 26*, 361–369.

10. Malika, M., & Shriram, S (2020). Effect of nanoparticle mixed ratio on stability and thermo-physical properties of CuO-ZnO/water-based hybrid nanofluid. *Journal of the Indian Chemical Society, 97*(3), 414–419.

11. Malika, M., Ashokkumar, M., & Sonawane, S. S. (2022). Mathematical and numerical investigations of nanofluid applications in the industrial heat exchangers. In *Applications of Nanofluids in Chemical and Bio-medical Process Industry* (pp. 53–78). Amsterdam: Elsiver. https://doi.org/10.1016/B978-0-323-90564-0.00010-6

6 Enhancement of Heat Transfer Using Nanofluids

6.1 INTRODUCTION TO HEAT TRANSFER

Heat transfer is the phenomenon in which thermal energy is transferred from a higher temperature object or area to a lower temperature object or area. The motion is propelled by the disparity in temperature between the items or areas. Conduction, convection, and radiation are the primary mechanisms responsible for heat transfer (Figure 6.1). These mechanisms are essential in a wide range of applications and industries. Gaining a comprehensive understanding of these mechanisms and their practical uses is crucial for maximizing operational processes, enhancing productivity, and guaranteeing safety in various industries.

6.1.1 MECHANISMS OF HEAT TRANSFER

6.1.1.1 Conduction

Conduction refers to the process by which heat is transferred through a solid substance, as individual particles pass on thermal energy to neighboring particles without any overall movement of the material. It happens when particles within a substance collide, resulting in the transfer of kinetic energy. Metals exhibit excellent thermal conductivity as a result of the abundance of unbound electrons that efficiently transport thermal energy throughout the material. Fourier's Law governs the rate of heat transmission by conduction. It asserts that the rate of heat transfer is directly proportional to the negative temperature gradient and the area across which the heat flows.

6.1.1.2 Convection

Convection refers to the process of heat transmission through the movement of a fluid, which can be either a liquid or a gas. It can occur spontaneously or be induced. Natural convection arises from the influence of buoyancy, wherein warmer fluid with lower density ascends while colder fluid with higher density descends. Forced convection refers to the utilization of external mechanisms, such as fans or pumps, to induce the movement of the fluid. Convection is governed by Newton's Law of Cooling, which establishes a relationship between the rate of heat transfer and the temperature differential between the fluid and the surface, as well as the heat transfer coefficient.

DOI: 10.1201/9781003594949-6

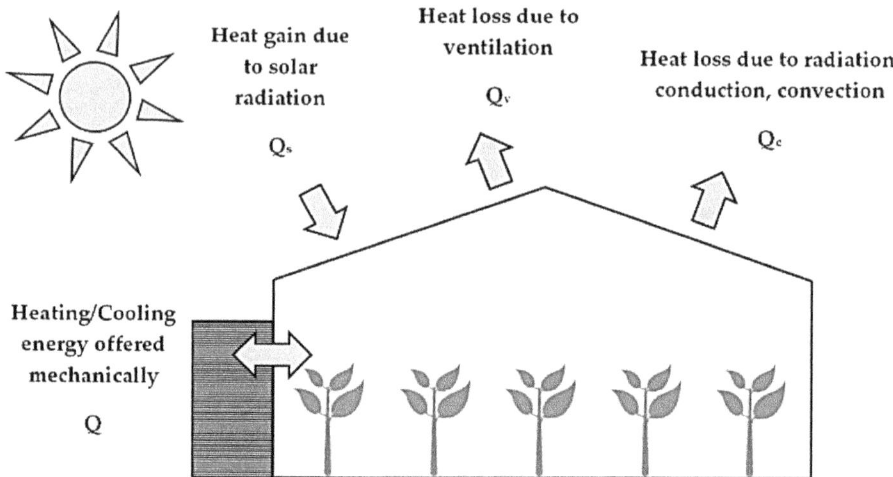

FIGURE 6.1 Illustration of the heat transfer mechanism in a standard greenhouse: Here, heat gain is due to solar radiation; the heat loss is due to radiation, conduction, and convection.[1]

6.1.1.3 Radiation

Radiation refers to the process of heat transmission by the emission of electromagnetic waves, specifically infrared radiation. Radiation, unlike conduction and convection, may proceed through a vacuum and does not rely on a medium. The emission and absorption of radiant energy by all objects is governed by the Stefan-Boltzmann law, which states that the quantity of energy radiated is directly proportional to the fourth power of the object's absolute temperature.

6.1.2 Importance of Heat Transfer in Different Industries

6.1.2.1 Electronics

The electronics industry places great importance on heat transfer as it is crucial for maintaining optimal performance and durability of gadgets. During operation, electronic components produce heat that needs to be dissipated in order to avoid overheating and potential malfunction. Heat sinks, thermal interface materials, and cooling fans are frequently employed to regulate the transfer of heat. High-performance applications, such as data centers and gaming computers, also utilize advanced cooling technologies, such as liquid cooling and heat pipes.

6.1.2.2 Building Ventilation and Cooling

Heating, Ventilation, and Air Conditioning (HVAC) systems utilize heat transfer principles to control indoor temperatures, guaranteeing both comfort and energy economy. During the heating mode, these systems facilitate the transmission of heat from a furnace or heat pump to the indoor air. During the cooling mode, they extract heat from the inside air and expel it outdoors. Efficient thermal insulation and architectural design are also vital in avoiding undesired heat transmission, decreasing energy usage, and ensuring a pleasant indoor atmosphere.

6.1.2.3 Aeronautics

Thermal management is crucial in the aeronautics business because of the extremely high temperatures experienced during flight and re-entry. Aircraft and spacecraft necessitate heat dissipation systems to manage the thermal energy produced by engines, electronics, and atmospheric friction. Thermal protection devices, including heat shields and insulating tiles, are crucial for safeguarding spacecraft during the process of re-entry. Radiators and heat exchangers are employed to regulate thermal energy in aviation and spacecraft systems, guaranteeing secure and effective functioning.

6.1.2.4 Transportation

The automotive sector heavily depends on efficient heat transfer to uphold engine performance and dependability. Combustion engines generate substantial quantities of heat that need to be dispersed in order to avoid excessive heating. Radiators, intercoolers, and oil coolers serve the purpose of dissipating heat from the engine. Battery temperature management systems are essential in electric and hybrid vehicles to ensure optimal battery performance and durability. These systems employ either liquid or air cooling to control the temperature of the battery.

6.1.2.5 Nuclear Reactors

Efficient heat transport is crucial for ensuring safety and optimizing performance in nuclear power plants. Nuclear reactors produce vast quantities of thermal energy through the process of nuclear fission. In order to prevent overheating and probable meltdowns, it is necessary to transport this heat away from the reactor core. Heat exchangers and cooling systems, such as main and secondary coolant loops, are specifically engineered to control and regulate the transfer of heat. The main coolant collects thermal energy from the reactor core and sends it to a secondary circuit, which in turn powers turbines to produce electricity.

6.1.2.6 Pharmaceuticals

The pharmaceutical sector necessitates meticulous temperature regulation throughout the manufacturing and preservation of drugs. Precise regulation of temperature is essential for preserving the stability and effectiveness of pharmaceutical goods. Heat transfer concepts are utilized in a range of processes, such as chemical reactors, crystallizers, and lyophilization (freeze-drying). Ensuring efficient heat transmission is essential for preserving the integrity of active substances and reaching the appropriate level of product quality.

6.1.2.7 Food Industry

The food sector heavily relies on heat transfer for crucial procedures such as cooking, pasteurization, freezing, and refrigeration, which are necessary to ensure food safety and maintain high quality. Heat exchangers are utilized for the effective transfer of heat in many procedures, such as pasteurization, which involves heating milk to eliminate detrimental germs. Efficient heat transfer is essential for ovens and cooking equipment to ensure the even cooking of food products. Heat extraction is necessary in refrigeration and freezing processes to maintain the freshness of food and avoid its deterioration. Comprehending and managing the passage of heat is crucial for maximizing these procedures and guaranteeing the safety of food.

6.1.2.8 Manufacturing Sector

In the field of manufacturing, numerous operations entail substantial heat transmission. Precise heat management is essential in casting, welding, forging, and machining processes to attain the desired material properties and ensure high product quality. During the casting process, it is necessary to carefully regulate the cooling of molten metal to get the correct mechanical qualities in the resulting solid pieces. Heat is employed in the process of welding to unite materials, and precise regulation of heat input is essential in order to prevent the occurrence of flaws. Novel thermal management methods, such as the utilization of nanofluids and phase change materials, are currently being devised to improve the efficiency of heat transfer in industrial procedures.

6.1.2.9 Renewable Energy

Renewable energy technologies utilize effective heat transfer to convert natural energy sources into usable forms. Solar thermal systems utilize heat transfer fluids to capture and transport heat from the sun in order to generate electricity or provide heating. Geothermal systems harness thermal energy from the Earth and transfer it to buildings for the purpose of heating and cooling. Efficient cooling of generators and power electronics is essential in wind turbines to ensure optimal performance. Heat transfer plays a crucial role in bioenergy production, where energy is generated from biomass by means of processes such as combustion and gasification.

6.1.2.10 Chemical Processing

The chemical processing industry relies on efficient heat transmission to regulate reaction temperatures and isolate chemical components. Heat exchangers facilitate the transmission of thermal energy between different process streams, ensuring that the temperatures required for chemical reactions are maintained at their ideal levels. Distillation columns utilize heat transfer to separate mixtures by exploiting variations in boiling points. Controlling heat transmission is crucial in reactors to attain the optimum reaction rates and product yields. Efforts are underway to investigate advanced heat transfer technologies, such as microchannel reactors and nanofluids, in order to improve the efficiency and performance of chemical processing.

6.2 HEAT TRANSFER ENHANCEMENT TECHNIQUES

Heat transfer enhancement techniques refer to a range of approaches and technologies that aim to enhance the efficiency and efficacy of heat transfer processes. These strategies are vital in diverse industries where effective heat transfer is necessary for optimal performance, energy conservation, and safety. Below are many prominent methods for enhancing heat transfer:

6.2.1 Surface Modification Techniques

- **Fins** refer to elongated structures that are affixed to heat transfer surfaces, such as pipes or walls, with the purpose of augmenting the surface area available for convection and radiation. They enhance heat transfer rates by augmenting the contact area between the surface and the adjacent fluid.
- **Ribs and grooves** refer to surface patterns or features that are incorporated into heat transfer surfaces in order to increase turbulence and facilitate the mixing of fluids. They cause disturbances in the smooth flow of a fluid, which in turn leads to an increase in the transmission of heat by convection.
- **Roughening the surface** can improve convective heat transfer by inducing turbulence at the boundary layer and increasing the effective surface area for heat exchange.

6.2.2 High-Level Heat Transfer Fluids

- **Nanofluids** refer to deliberately created suspensions of nanoparticles, such as metals, oxides, or carbon-based compounds, within a base fluid, usually water or oil. They demonstrate a substantial increase in thermal conductivity compared to traditional fluids, therefore enhancing the efficiency of heat transmission in many applications.
- **Phase change materials** are substances that have the ability to absorb and release thermal energy when they undergo a phase transition, such as changing from a solid to a liquid or vice versa. These devices are utilized for the purpose of storing and releasing significant quantities of latent heat, hence successfully controlling temperatures in various applications such as thermal energy storage, building insulation, and electronics cooling.

6.2.3 Modifications to the Design of the Heat Exchanger

- *Microchannel heat exchangers*: Microchannels are channels that have small dimensions, usually with hydraulic diameters smaller than 1 mm. They improve convective heat transfer by increasing the ratio of surface area to volume and by allowing fluid to flow at faster velocities. Compact heat exchangers utilize them to achieve efficient heat transfer in confined areas.
- Compact heat exchangers employ sophisticated designs, such as plate-fin layouts, to optimize surface area and enhance heat transfer efficiency. These devices are frequently employed in HVAC systems, refrigeration, and aeronautical applications.

6.2.4 Techniques for Advanced Thermal Management

- **Heat pipes** are passive heat transfer devices that effectively convey heat over long distances by utilizing phase change principles. These devices are composed of a hermetically sealed tube that contains a working fluid. The fluid evaporates at the heat source, moves to the cool end through vapor pressure, condenses, and then returns to the heat source through capillary action.

- **Thermoelectric coolers** employ the Peltier effect to establish a heat transfer mechanism by means of the electrical current passing between two distinct materials. They are capable of providing accurate temperature regulation in electronic devices and small-scale cooling tasks.

6.2.5　Enhanced Heat Transfer in Fluid Flow

- *Improved heat transmission in fluid flow turbulators*: Turbulators are devices that are introduced into the routes of fluid flow, such as tubes or ducts, in order to create turbulence and increase the efficiency of convective heat transmission. They disturb the smooth flow of a fluid, which leads to higher rates of heat transmission by encouraging the mixing of the fluid and diminishing the layers of heat near the surface.
- Jet impingement is the process of sending high-velocity jets of fluid onto a surface in order to improve convective heat transfer. It enhances the efficiency of heat transport by disrupting sluggish boundary layers and facilitating the mixing process.

6.2.6　Enhancement of Radiative Heat Transfer

- Selective surfaces possess a high degree of emissivity toward infrared radiation, allowing them to efficiently emit heat, while simultaneously exhibiting low absorptivity toward solar radiation, hence limiting heat gain. Solar thermal collectors and passive solar heating systems utilize them.
- Radiative shields are designed to deflect thermal radiation, thereby minimizing heat transfer through radiation. Insulation materials and spacecraft thermal management systems utilize them.

The utilization of nanofluids as coolants and heat transfer fluids has acquired significant advantage among these strategies. The next sections will cover the improvement of heat transfer through the utilization of nanofluids.

6.3　ENHANCEMENT OF HEAT TRANSFER USING NANOFLUIDS

The improvement of convective heat transfer and thermal conductivity of liquids was initiated by incorporating micron-sized particles into a base fluid. Nevertheless, the method has remained impractical due to the quick accumulation of sediment, erosion, blockage, and the substantial decrease in pressure generated by these particles. A minute quantity of nanoparticles, when evenly distributed and securely suspended in base fluids, can yield remarkable enhancements in the thermal characteristics of those fluids.

Nanofluids are colloidal mixtures of nanoparticles (1–100 nm) and a base liquid. The term was coined by Choi[2] in 1995 at the Argonne National Laboratory to describe a new class of heat transfer fluids based on nanotechnology. These fluids have thermal properties that are superior to those of their base fluids or conventional particle fluid suspensions.

Aim: The objective of nanofluids is to optimize thermophysical properties while minimizing the solid volume fraction (weight % <1%) of nanoparticles in the base fluids. Therefore, the crucial factor for achieving a substantial improvement in heat transfer is the suspension of nanoparticles that are nearly non-agglomerated or mono-dispersed in liquids.

6.3.1 IMPORTANT THERMOPHYSICAL PROPERTIES OF HEAT TRANSFER FLUIDS

The thermophysical properties of the base fluid are essential in determining and improving the heat transfer properties of nanofluids. The characteristics of the nanoparticles interact with the nanofluids, which affects their overall thermal performance. Important thermophysical properties of conventional heat transfer fluids are illustrated in Figure 6.2. The following are the crucial thermophysical parameters of the base fluid that enhance the heat transmission characteristics in nanofluids.

6.3.1.1 Thermal Conductivity

Thermal conductivity is the most crucial characteristic that influences the efficiency of heat transmission in nanofluids. The presence of nanoparticles in a base fluid with higher thermal conductivity improves heat conduction within the fluid, resulting in increased overall heat transfer rates.

Enhancement mechanism: The addition of nanoparticles to the base fluid can greatly enhance its heat conductivity, surpassing that of the pure fluid. The improvement is impacted by the nanoparticle's composition, concentration, and distribution quality.

6.3.1.2 Viscosity

Viscosity plays a crucial role in determining the properties of fluid flow and the efficiency of heat transfer. Fluids with lower viscosity typically exhibit decreased resistance to flow, which enhances convective heat transfer. Nevertheless, an overly low viscosity can result in difficulties in maintaining the stability of nanoparticle suspensions.

Enhancement mechanism: The enhancement mechanism involves the ability of nanoparticles to modify the viscosity of the base fluid, which is influenced by factors such as their size, shape, and concentration. Strategic choice and distribution of nanoparticles can reduce viscosity enhancements, guaranteeing efficient fluid flow and heat transfer efficiency.

6.3.1.3 Density

The density of a fluid affects the convection induced by buoyancy and the pressure drop in flow systems. Decreased densities can improve the transport of heat by

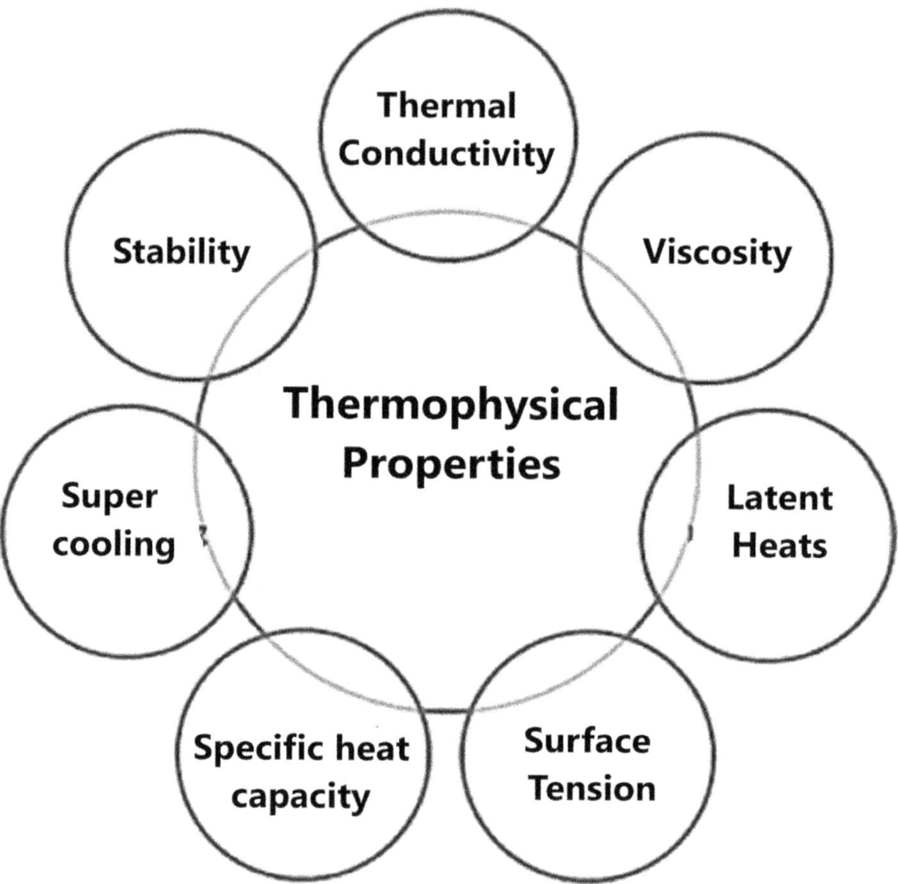

FIGURE 6.2 Important thermophysical properties of conventional base fluids for the potential heat transfer application.[3]

natural convection, but increased densities can amplify the influence of gravity in larger systems.

Enhancement mechanism: The presence of nanoparticles in a fluid has minimal impact on its density since the nanoparticles are very small and exist in low quantities compared to the main fluid. Therefore, nanofluids generally experience modest changes in density, which allows them to maintain fluid behavior that is similar to the base fluid.

6.3.1.4 Specific Heat Capacity

Specific heat capacity is crucial as it quantifies the quantity of thermal energy needed to alter the temperature of a specified amount of fluid. Fluids with a higher specific heat capacity have the ability to store a greater amount of thermal energy in a given volume. This property helps to maintain a constant temperature and provides thermal inertia.

Enhancement mechanism: Nanoparticles have the potential to modestly modify the specific heat capacity of the base fluid, based on their material qualities and concentration. The modification of this property is typically negligible and does not have a substantial impact on the efficiency of heat transfer in nanofluids.

6.3.1.5 Boiling and Freezing Points

Boiling and freezing points determine the temperature ranges at which operations can be carried out and ensure stability in applications involving heat transmission. Fluids with elevated boiling points are appropriate for operations conducted at high temperatures, but lower freezing points hinder the solidification of fluids at low temperatures.

Enhancement mechanism: The use of nanoparticles can modify the boiling and freezing points of the base fluid, hence impacting the efficiency of heat transfer under extremely hot or cold situations. Surface changes or the use of surfactants can alleviate these effects, guaranteeing consistent fluid behavior across a broad spectrum of temperatures.

6.3.1.6 Chemical Compatibility and Stability

Chemical compatibility and stability are crucial factors that guarantee the long-term effectiveness and dependability of nanofluids under different operational circumstances. Ensuring heat transfer efficiency requires the compatibility of system materials and their resistance to deterioration or corrosion.

Enhancement mechanism: In order to maintain the stability and performance of nanofluids, it is crucial that the nanoparticles and base fluids are compatible with each other to avoid any chemical reactions or particle aggregation that could lead to degradation. Chemical compatibility and stability can be improved through surface functionalization or the appropriate application of surfactants.

6.3.2 ENHANCEMENT OF THERMAL CONDUCTIVITY

The addition of nanoparticles to a base fluid facilitates various complex mechanisms that boost thermal conductivity in nanofluids as shown in Figure 6.3. Ultrasonication is an essential process parameter utilized in the manufacturing of nanofluids. It plays a critical role in dispersing nanoparticles and ultimately impacts the enhancement of thermal conductivity. The augmentation is determined by various factors, such as the thermal conductivity of both the base fluid and nanoparticles, pH level, stability, temperature, kind of nanoparticles, their size, concentration, and shape. Let us thoroughly examine each parameter, providing examples, in order to comprehend their influence on the thermal conductivity of nanofluids.

6.3.2.1 Thermal Conductivity of Base Fluid and Nanoparticles

6.3.2.1.1 Base Fluid

The base fluid functions as the medium for the dispersion of nanoparticles. Typical base fluids consist of water, ethylene glycol, oils, and refrigerants. The inherent thermal conductivity of the base fluid establishes the fundamental level for the thermal efficiency of the nanofluid. Water is commonly employed as a base fluid because it

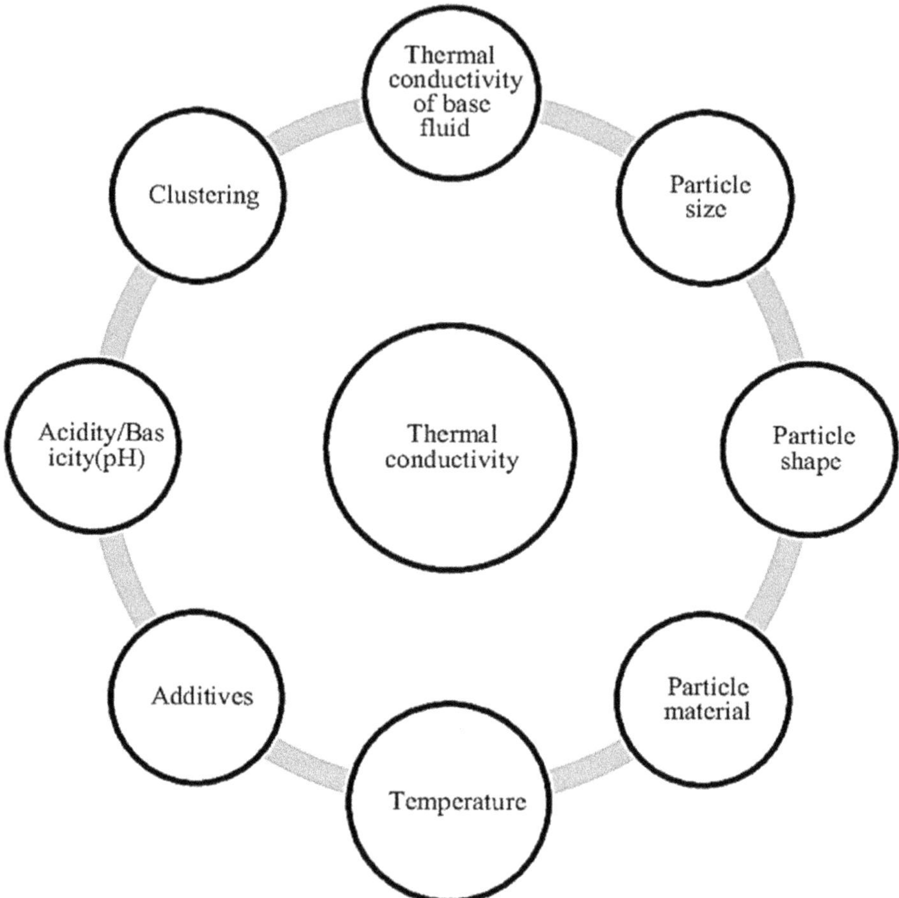

FIGURE 6.3 Main parameters affecting the thermal conductivity of nanofluids.[4]

has a thermal conductivity of about 0.6 W/mK, which is quite high and readily available. On the other hand, ethylene glycol exhibits a reduced thermal conductivity of around 0.25 W/mK. However, it is commonly employed in automotive cooling systems because of its ability to prevent freezing.

6.3.2.1.2 Nanoparticles

Nanoparticles have significantly greater thermal conductivities in comparison with the basic fluids. Typical examples of nanoparticles include metallic elements like copper and silver, as well as metal compounds like alumina and silica. For example, copper nanoparticles exhibit a thermal conductivity of around 400 W/mK, which is considerably more than that of water. The choice of nanoparticle can significantly impact the overall thermal conductivity of the nanofluid. For instance, the introduction of copper nanoparticles into water can increase its thermal conductivity by more than 40% at specific concentrations.

6.3.2.2 Nanofluid's pH Level

The pH level of the nanofluid has an impact on the surface charge of nanoparticles, which in turn influences their stability and dispersion. Ensuring an adequate pH level is crucial for maintaining the dispersion of nanoparticles and preventing their agglomeration. Aggregation results in the formation of bigger particle clusters which then settle down, causing a decrease in the overall thermal conductivity. The ideal pH level changes depending on the specific nanoparticle substance. Alumina nanoparticles in water exhibit the most stability and dispersion when the pH is approximately 8–9, which improves the thermal conductivity of the nanofluid.

6.3.2.3 Nanofluid Stability

Consistent thermal performance relies heavily on stability. Over time, nanofluids have a tendency to form agglomerates, which results in sedimentation and a decrease in thermal conductivity. Enhancing stability can be achieved by employing surfactants or stabilizers, as well as employing mechanical techniques such as ultrasonication.

6.3.2.3.1 Surfactants and Stabilizers

Surfactants and stabilizers, such as cetyltrimethylammonium bromide (CTAB) or sodium dodecyl sulfate (SDS), are additives that prevent the clumping of nanoparticles by creating a barrier based on steric hindrance or electrostatic repulsion.

Ultrasonication is a mechanical procedure that evenly distributes nanoparticles in a fluid, separating clumps and improving stability. Ultrasonication has been demonstrated to greatly enhance the stability and thermal conductivity of alumina-water nanofluids.

6.3.2.4 Nanofluid Temperature

The thermal conductivity of nanofluids is greatly influenced by temperature. In general, the thermal conductivity of nanofluids increases as the temperature rises, mostly because the nanoparticles experience greater Brownian motion. The heightened movement results in enhanced thermal exchanges between the nanoparticles and the base fluid.

Example: A study showed that the thermal conductivity of a nanofluid composed of ethylene glycol and copper nanoparticles improved by almost 30% when the temperature was elevated from 25°C to 50°C. The impact of this phenomenon is contingent upon the specific nanoparticle and base fluid utilized; nevertheless, it is commonly noted that thermal conductivity tends to increase as temperature rises.

6.3.2.5 Nanoparticle Type

Nanoparticles of different sorts exhibit varied degrees of augmentation in heat conductivity, depending on their material qualities.

- Metal nanoparticles, such as copper, silver, and gold, possess excellent thermal conductivity; yet, they can be costly and susceptible to oxidation.
- Metal oxide nanoparticles, such as alumina, silica, and titania, are characterized by greater stability and ease of handling. However, their

thermal conductivity increases are often lower compared to metals. Alumina nanoparticles dispersed in water can increase thermal conductivity by approximately 20%–30%, but copper nanoparticles can increase it by more than 40%.

6.3.2.6 Nanoparticle Size

The dimensions of nanoparticles directly affect their surface area and their interaction with the base fluid. Smaller nanoparticles exhibit a higher ratio of surface area to volume, which can augment thermal conductivity but also result in heightened aggregation.

Example: A study discovered that nanofluids consisting of water and alumina nanoparticles measuring 20 nm shown a greater improvement in thermal conductivity compared to those containing 50-nm particles. Nevertheless, nanoparticles that are excessively tiny can result in substantial agglomeration, hence diminishing the efficacy of the nanofluid.

6.3.2.7 Nanoparticle Concentration

The nanoparticle concentration in the base fluid is a crucial determinant of thermal conductivity.

- Higher concentration leads to an increase in thermal conductivity because it results in a greater number of conductive channels. For instance, the addition of copper nanoparticles to water in a volume ratio of 5% can enhance its thermal conductivity by roughly 50%.
- *Optimal concentration*: Beyond a certain concentration, the thermal conductivity may not increase considerably and can result in problems such as higher viscosity and sedimentation.

Example: For instance, a study demonstrated that if the concentration of silica nanoparticles in water exceeds 3% by volume, the improvement in thermal conductivity levels out and the viscosity increases significantly, resulting in problems with pumping and flow.

6.3.2.8 Influence of Nanoparticle Shape

The morphology of nanoparticles influences their interactions and thermal conductivity.

- *Comparison of spherical and non-spherical nanoparticles*: Non-spherical nanoparticles, such as rods or platelets, exhibit enhanced heat conductivity as a result of their increased aspect ratio. Carbon nanotubes (CNTs) have a remarkably high thermal conductivity and can greatly improve the thermal characteristics of the base fluid. A study discovered that the incorporation of CNTs into water resulted in a significant enhancement of its thermal conductivity, exceeding 50% at low concentrations. This improvement can be attributed to the exceptional aspect ratio and thermal conductivity properties exhibited by CNTs.

- *Shape anisotropy*: The presence of shape anisotropy in the nanofluid can result in the formation of more effective routes for heat transfer, hence improving the augmentation of thermal conductivity. Platelet-shaped nanoparticles have demonstrated favorable outcomes in augmenting thermal conductivity in comparison with spherical nanoparticles.

6.3.2.9 Interaction between Nanoparticles and Base Fluid

The interplay between nanoparticles and the underlying fluid can have a substantial impact on thermal conductivity.

- Surface functionalization involves modifying the surface of nanoparticles to enhance their compatibility and dispersion in the base fluid. Surface-modified silica nanoparticles containing hydrophilic groups improve the dispersion in water, resulting in enhanced thermal conductivity.
- Reducing the interfacial thermal resistance between the nanoparticles and base fluid is essential for optimizing thermal conductivity. To decrease this resistance, one can employ efficient surface treatments and select nanoparticle-base fluid combinations that are compatible with each other.

6.3.3 Effect of Ultrasonication on Enhancing Thermal Conductivity of Nanofluids

Ultrasonication is a commonly employed method for increasing the thermal conductivity of nanofluids. The procedure entails utilizing ultrasonic waves to evenly distribute nanoparticles throughout a base fluid, disrupting clusters and facilitating the formation of stable suspensions. This section provides a comprehensive analysis of how ultrasonication impacts the thermal conductivity of nanofluids.

6.3.3.1 Mechanism of Ultrasonication

The process of ultrasonication involves the use of high-frequency sound waves to induce mechanical vibrations in a substance.

Ultrasonication creates high-frequency sound waves that induce alternating cycles of high pressure and low pressure in the liquid medium. Cavitation is the result of cycles that cause the development of tiny bubbles that expand and burst forcefully. Cavitation generates extreme local conditions, characterized by elevated temperatures and pressures, which facilitate the disintegration of nanoparticle agglomerates and promote a homogeneous dispersion.

6.3.3.2 Effect on Thermal Conductivity Enhanced Dispersion

- Ultrasonication achieves a more uniform dispersion of nanoparticles inside the base fluid.
- Uniform dispersion inhibits the development of nanoparticle clusters, which can impede heat conductivity by generating zones of thermal resistance.
- Improved dispersion facilitates enhanced interaction between the nanoparticles and the base fluid, hence facilitating more efficient heat transfer pathways as shown in Figure 6.4.

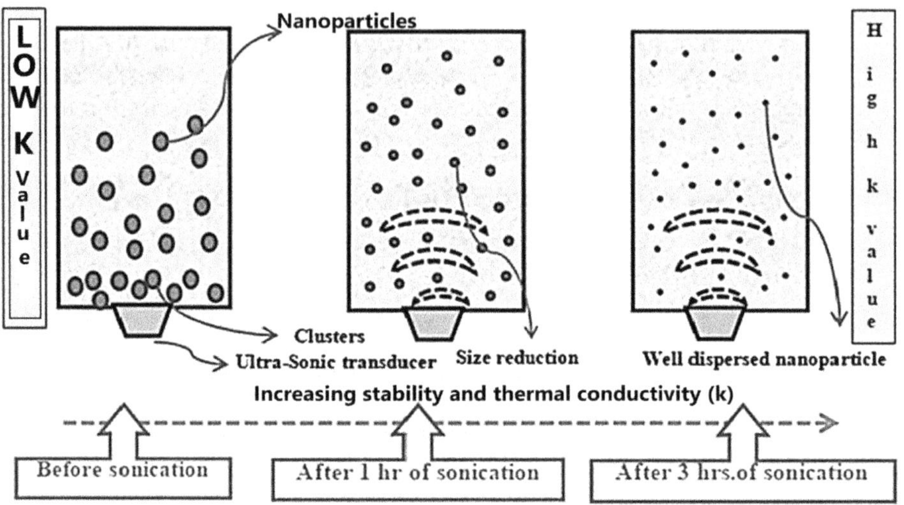

FIGURE 6.4   Effect of ultrasonication on the stability and thermal conductivity of nanofluids.[5]

Important points to remember

i. **Enhanced Surface Area**

Nanomaterials, such as metal oxides, nanoclays, or carbon nanotubes, have a tendency to form agglomerates when they are introduced into a liquid. Efficient methods for breaking up and spreading apart the agglomerated particles are necessary to counteract the cohesive forces that occur after moistening the powder. Utilizing ultrasonic waves to break up agglomeration formed in both aqueous and nonaqueous solutions enables the complete exploitation of the potential of nanosized materials. Experiments conducted on different mixtures of nanoparticle clusters with varying amounts of solid material have shown that ultrasound is significantly more effective than other methods, such as rotor stator mixers (like ultra turrax), piston homogenizers, or wet milling techniques (such as bead mills or colloid mills). Hielscher ultrasonic systems can operate at rather high levels of solid concentrations. For instance, in the case of silica, it was discovered that the rate at which it breaks does not depend on the concentration of the solid material, as long as it is below 50% by weight. Ultrasound can be used to disperse high concentration master-batches, which involves processing liquids with both low and high viscosity. Ultrasound is an effective processing method for paints and coatings, as it can be used with many media like water, resin, and oil. Dispersing nanoparticle agglomerates enhances the nanoparticles' effective surface area, hence facilitating heat transfer as shown in Figure 6.5. Increased surface area facilitates enhanced thermal energy transfer between the nanoparticles and the fluid.

ii. **Decreased Thermal Resistance**

Ultrasonication decreases the overall heat resistance in the nanofluid by stopping the production of big clumps. Reduced particle size and even distribution enhance the efficiency of heat transfer.

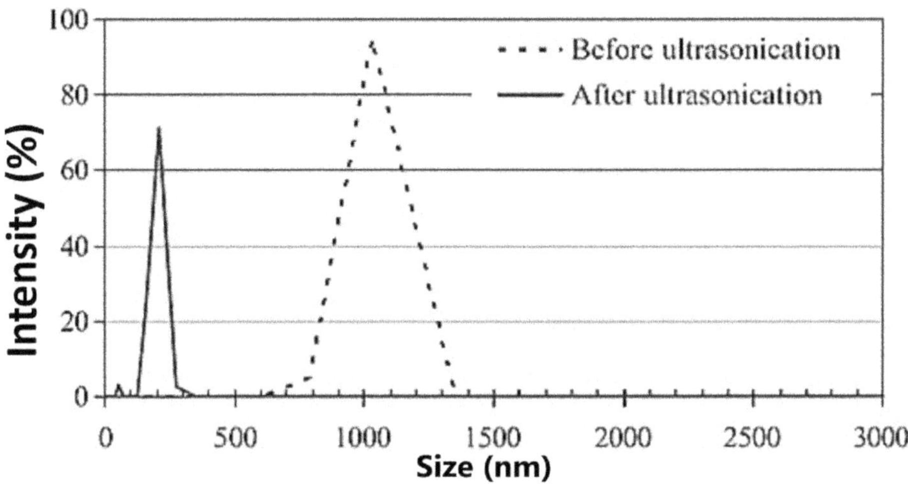

FIGURE 6.5 Effect of ultrasonication on the deagglomeration of nanoparticles in nanofluids.

Variables Affecting the Efficiency of Ultrasonication

iii. Ultrasonication Time

- Extended ultrasonication can result in improved dispersion at first, but excessive ultrasonication might potentially lead to re-agglomeration or harm to the nanoparticles.
- The optimal duration for ultrasonication varies depending on the specific type and concentration of nanoparticles, as well as the parameters of the base fluid.

iv. Ultrasonic Power and Frequency

Ultrasonic Power

- *Influence on cavitation intensity*

  The intensity of cavitation, which refers to the development and collapse of small bubbles in the fluid, is determined by the ultrasonic power. Increased power levels enhance cavitation intensity, resulting in more potent shear forces and shock waves that effectively disintegrate nanoparticle agglomerates.

- *Dispersion efficiency*

  When power levels are at their best, the heightened intensity of cavitation leads to improved dispersion of nanoparticles, resulting in a decrease in particle size and a more even distribution as shown in Figure 6.6. This facilitates the attainment of a stable nanofluid with improved thermal characteristics.

- *Effects of overpowering:*
  - Re-agglomeration can occur when nanoparticles are subjected to excessive ultrasonic power, resulting in the excessive input of energy. This might lead to instability in the suspension.

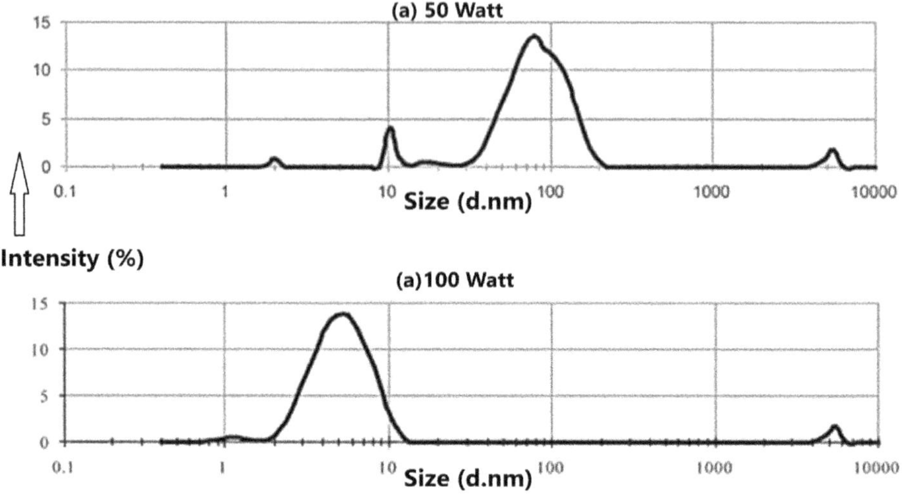

FIGURE 6.6   (a) 50 W and (b) 100 W: effect of ultrasonication power on the particle size distribution of nanofluids.[6]

- *Heat generation*: The production of intense ultrasonic power results in substantial heat generation, which can cause the deterioration of both the base fluid and the nanoparticles. This, in turn, has an impact on the stability and effectiveness of the nanofluid.

Ultrasonic Frequency

Ultrasonic frequency refers to the frequency of sound waves that are higher than the upper limit of human hearing, often above 20,000 Hz.

- *Frequency and cavitation*
  - *Low frequency (20–40 kHz)*: Ultrasonication at low frequencies generates larger cavitation bubbles that collapse with greater intensity, resulting in the production of powerful shear forces. This process is efficient in disintegrating bigger clusters and decreasing the size of particles. Nevertheless, it can also result in a more substantial production of heat.
  - *High frequency (40–200 kHz and above)*: Ultrasonication at high frequencies produces smaller cavitation bubbles that collapse with less intensity but at a higher pace. This leads to a more precise distribution and a more consistent decrease in particle size, with a decreased likelihood of excessive heating.
- *Optimal frequency range*: The ideal frequency is contingent upon the specific characteristics of the nanoparticles and the intended particle size. For instance, metal oxide nanoparticles may necessitate distinct frequencies in comparison with carbon-based nanoparticles in order to attain the most effective dispersion and reduction in size.

Summary

Increasing the intensity of ultrasonic waves and using the optimal frequency can amplify the formation of cavitation, resulting in improved dispersion. Nevertheless, an excessively high power level has the potential to generate an excessive amount of heat, which could have a detrimental impact on the stability of the nanofluid.

v. Type of Nanoparticles

Nanoparticles exhibit varied responses to ultrasonication depending on their material qualities, size, and shape. Optimal dispersion of metal oxide nanoparticles, carbon-based nanoparticles, and other types of nanoparticles may necessitate the use of customized ultrasonication parameters.

vi. Ultrasonic Transducer Type

The type of ultrasonic transducer employed in the production of nanofluids has a substantial influence on their thermal conductivity. Various transducers, including basic magnetic stirrers, bath-type sonicators, and probe-type sonicators, exhibit different degrees of efficacy in dispersing nanoparticles and thereby improving thermal conductivity. This section provides a comprehensive examination of each category of ultrasonic transducer and its impact on the thermal conductivity of nanofluids.

Mechanism of a Simple Magnetic Stirrer

A magnetic stirrer employs a revolving magnetic field to rotate a stir bar submerged in the liquid, generating a whirlpool that facilitates blending. Efficiency:

- *Restricted dispersive capacity*: Although magnetic stirrers are capable of blending nanoparticles into the base fluid, they are often less efficient in disintegrating nanoparticle agglomerates compared to ultrasonic approaches.
- *Particle size*: Magnetic stirring typically leads to a larger and less uniform particle size distribution, which hinders effective heat transmission pathways within the nanofluid.
- The thermal conductivity is generally reduced due to the existence of bigger agglomerates and uneven dispersion, resulting in a lower enhancement.

Application:

- Appropriate for initial blending or fluids with low thickness when there is no need for intense shear pressures.

Mechanism of a Bath-Type Sonicator

A bath-type sonicator utilizes ultrasonic waves produced in a water bath to transfer energy into the container holding the sample placed in the bath. It offers consistent energy distribution, albeit at a reduced intensity compared to probe sonicators. Efficiency:

- Bath sonicators have a higher capacity for dispersion compared to magnetic stirrers due to their ability to create cavitation throughout the bath. Nevertheless, the energy intensity is reduced in comparison with probe sonicators.

- *Particle size*: This technique produces nanoparticles that are smaller and more evenly dispersed compared to magnetic stirring, although they are not as tiny as those achieved using probe sonication.
- The thermal conductivity shows a moderate increase as a result of better, although not ideal, dispersion of nanoparticles.

Application:

- This equipment is well suited for handling huge quantities and delicate samples that require prevention of concentrated heat and strong mechanical stress.

Mechanism of a Probe-Type Sonicator

A probe-type sonicator submerges an ultrasonic probe with high intensity directly into the nanofluid. The probe creates highly concentrated cavitation, resulting in strong shear forces that dismantle clusters of nanoparticles.[6]

Efficiency:

- Probe sonicators have exceptional dispersive power, enabling them to efficiently disintegrate agglomerates and achieve a consistent dispersion of nanoparticles.
- The approach employed in this process generates the lowest and most uniform distribution of particle sizes, resulting in a significant improvement in thermal conductivity.
- The heat conductivity is significantly enhanced as a result of the nanoparticles being dispersed optimally and suspended stably.[7]

Application:

- Most suitable for applications that need high accuracy and tiny amounts of samples, where attaining optimal dispersion and thermal performance is crucial.

Summary of Comparison Quality of Dispersion

- A magnetic stirrer is used for basic dispersion and is more effective for larger particle sizes, although it does not significantly boost thermal conductivity.
- The bath-type sonicator provides a moderate level of dispersion, resulting in intermediate particle size. It also offers a moderate improvement in thermal conductivity.
- A probe-type sonicator offers superior dispersion, resulting in the smallest particle size and the maximum boost in thermal conductivity.[8]

Improving Thermal Conductivity

- The effectiveness of the magnetic stirrer is restricted due to inadequate dispersion and the presence of bigger agglomerates.
- The bath-type sonicator shows some improvement in dispersion, but it is still considered moderate due to the fact that it has not achieved ideal dispersion.[9]

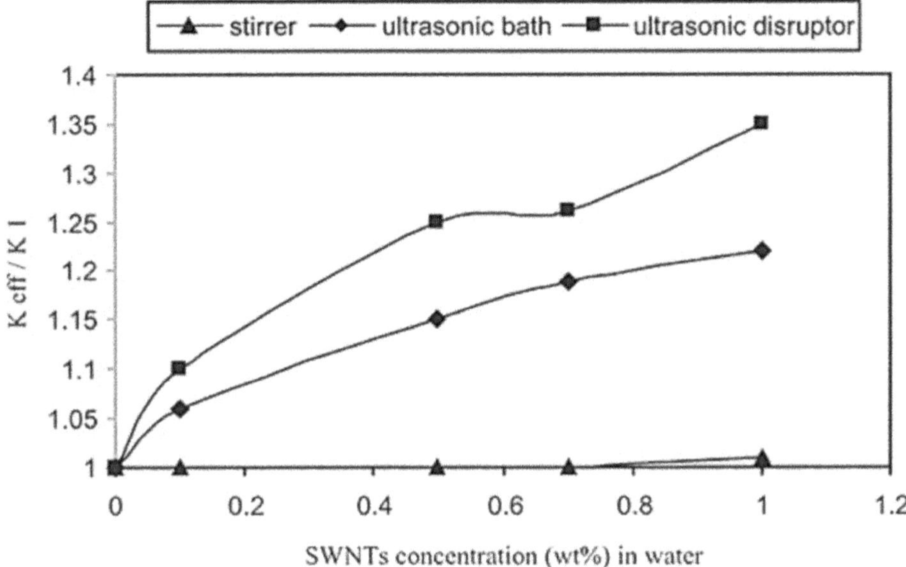

FIGURE 6.7 Effect of transducer type on the thermal conductivity of Single walled carbon nanotubes (SWCNT) nanofluids.[10]

- The probe-type sonicator achieves maximum performance by ensuring excellent dispersion and reducing particle size to a lower scale as shown in Figure 6.7.

  Compatibility
- The magnetic stirrer is designed for the purpose of initial mixing or for fluids with low viscosity.
- The bath-type sonicator is designed for processing bigger quantities and delicate samples.
- The probe-type sonicator is ideal for high-precision applications and small volumes that require maximum thermal performance.

6.4 SUMMARY

- Ultrasonication is a crucial method for developing nanofluids.
- It greatly improves their ability to conduct heat by ensuring that the nanoparticles are evenly dispersed and reducing resistance to heat flow.
- By fine-tuning the parameters of ultrasonication, the thermal efficiency of nanofluids can be maximized for a wide range of industrial and technological applications.
- The selection of the ultrasonic transducer significantly influences the thermal conductivity of nanofluids.
- Probe-type sonicators offer the greatest improvement in heat conductivity since they excel at evenly dispersing nanoparticles and successfully reducing particle size.

REFERENCES

1. Dimitropoulou, A.-M. N., Maroulis, V. Z., & Giannini, E. N. (2023). A simple and effective model for predicting the thermal energy requirements of greenhouses in Europe. *Energies, 16*(19), 6788. https://doi.org/10.3390/en16196788

2. Choi, S. U., & Eastman, J. A. (1995). *Enhancing Thermal Conductivity of Fluids with Nanoparticles* (No. ANL/MSD/CP-84938; CONF-951135-29). Argonne National Lab: Argonne, IL.

3. Munyalo, J. M., & Zhang, X. (2018). Particle size effect on thermophysical properties of nanofluid and nanofluid based phase change materials: A review. *Journal of Molecular Liquids, 265,* 77–87.

4. Ullah, A., Kilic, M., Habib, G. et al. (2023). Reliable prediction of thermophysical properties of nanofluids for enhanced heat transfer in process industry: a perspective on bridging the gap between experiments, CFD and machine learning. *Journal of Thermal Analysis and Calorimetry, 148,* 5859–5881. https://doi.org/10.1007/s10973-023-12083-7

5. Mukherjee, S., Mishra, P. C., Chakrabarty, S. et al. (2022). Effects of sonication period on colloidal stability and thermal conductivity of SiO_2–water nanofluid: an experimental investigation. *Journal of Cluster Science, 33,* 1763–1771. https://doi.org/10.1007/s10876-021-02100-w

6. Malika, M., Rao, C. V., Das, R. K., Giri, A. S., & Golder, A. K. (2016). Evaluation of bimetal doped TiO2 in dye fragmentation and its comparison to mono-metal doped and bare catalysts. *Applied Surface Science, 368,* 316–324.

7. Thakur, P., Pargaonkar, A., & Sonawane, S. S. (2023). Introduction to nanofluids. In *Nanofluid Applications for Advanced Thermal Solutions* (pp. 1–19). Elsevier: Amsterdam. https://doi.org/10.1016/B978-0-443-15239-9.00001-1

8. Thakare, Y., Dharaskar, S., Unnarkat, A., & Sonawane, S. S. (2022). Nanofluid-based drug delivery systems. In *Applications of Nanofluids in Chemical and Bio-medical Process Industry* (pp. 303–334). Elsevier: Amsterdam. https://doi.org/10.1016/B978-0-323-90564-0.00005-2

9. Thakur, P., Kumar, N., & Sonawane, S. S. (2021). Enhancement of pool boiling performance using MWCNT based nanofluids: A sustainable method for the wastewater and incinerator heat recovery. *Sustainable Energy Technologies and Assessments, 45,* 101115.

10. Amrollahi, A., Rashidi, A. M., Emami Meibodi, M., & Kashefi, K. (2009).. Conduction heat transfer characteristics and dispersion behaviour of carbon nanofluids as a function of different parameters. *Journal of Experimental Nanoscience, 4*(4), 347–363. https://doi.org/10.1080/17458080902929929.

7 Direct and Indirect Thermal Applications of Hydrodynamic and Acoustic Cavitation

7.1 INTRODUCTION TO HYDRODYNAMIC AND ACOUSTIC CAVITATION

Hydrodynamic and acoustic cavitation are separate occurrences that encompass the creation, enlargement, and implosion of bubbles within a liquid. Below is a comprehensive explanation and comparison of the two, accompanied by figures to enhance comprehension.

7.1.1 HYDRODYNAMIC CAVITATION: DEFINITION

Hydrodynamic cavitation is the phenomenon that occurs when the pressure within a fluid drops below its vapor pressure, resulting in the creation of vapor bubbles. This phenomenon commonly occurs in regions of elevated fluid velocity, such as in the vicinity of pump impellers, valves, or abrupt bends.

7.1.1.1 Mechanism

- *Pressure drop*: The movement of fluid is impeded as it passes through a narrow passage, such as an aperture or a valve.
- *Velocity increase*: The fluid's velocity augments as it traverses the constriction owing to the principle of mass conservation.
- *Pressure decrease*: As per Bernoulli's principle, the pressure diminishes as the velocity augments.
- *Bubble formation*: When the pressure drops below the vapor pressure of the fluid, bubbles of vapor are created.
- *Bubble collapse*: When the fluid travels away from the narrow point, the pressure rises once more, leading to a forceful collapse of the bubbles.

7.1.1.2 Hydrodynamic Cavitation Reactors

Hydrodynamic cavitation occurs when a liquid flows through a small, restricted passage such as a venturi tube, orifice plate, or throttling valve. Cavitation is the occurrence of fluid velocity and pressure being related according to Bernoulli's equation. This equation is utilized to ascertain the characteristics of incompressible fluid flow via a constriction, as expressed by the following equation.

DOI: 10.1201/9781003594949-7

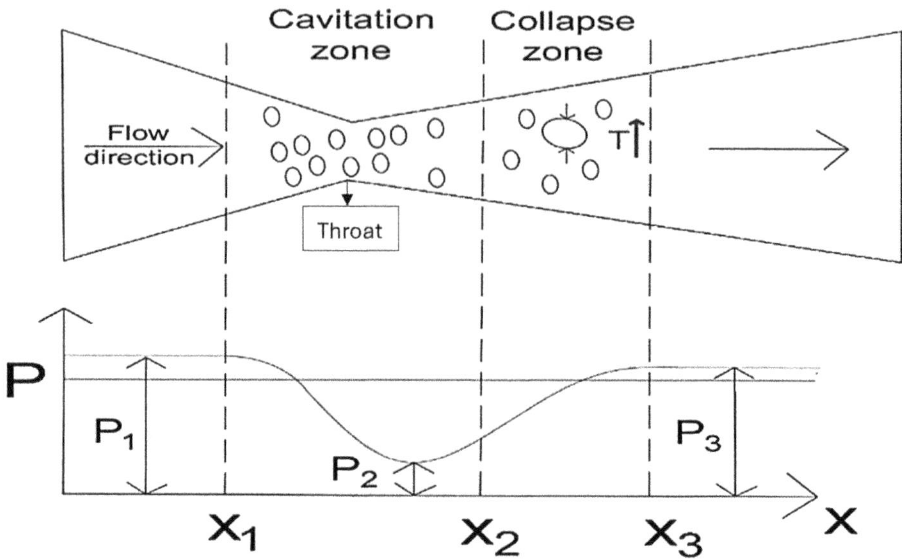

FIGURE 7.1   Schematic of venturi tube hydrodynamic reactor.[1]

$$P_1 + \frac{\rho V_1^{\,2}}{2} + gZ_1 = P_1 + \frac{\rho V_1^{\,2}}{2} + gZ_1 + \text{Energy of the fluid} \qquad (7.1)$$

Here, subscripts 1 and 2 represent the property in the throat section (Figure 7.1) and the pipe, respectively. Where, the properties P, V, and Z represent the pressure, fluid velocity, and height of the pipe, respectively. The energy of the fluid includes sum of static pressure, kinetic energy, potential energy, and the energy lost owing to resistance to flow.

This lost energy is converted into thermal energy as the fluid passes through the venturi. As fluid passes through a narrow tunnel, its velocity increases, and consequently, the pressure decreases. Cavities are created when the pressure in a specific region falls below the vapor pressure of the substance. As a result, the liquid expands in the direction of flow, leading to a recovery of pressure and subsequent collapse of the cavities. As a result, there is a formation of turbulent and violent airflow, accompanied by elevated pressure and temperature, which ultimately generates extremely damaging shock waves. When the amount of vapor exceeds the amount of liquid, shock waves will arise in void areas. The intensity of turbulence is determined by the magnitude of the pressure fall, the configuration of the constriction, and the rate at which pressure is restored. The cavitation intensity is determined by the maximal size of the bubbles. Due to a rapid implosion lasting only nanoseconds, the temperature inside the bubble rises to 5,000–10,000 K and the pressure reaches 1,000–2,000 atm. This high energy intensity, generated during the collapse of the bubble, creates the necessary conditions to break the agglomeration of the nanoparticles in the nanofluid. Hydrodynamic cavitation can be achieved by utilizing a rotational reactor, comprising a motor and a rotor and stator positioned in opposition to each other along the axial direction within the pump housing. The subsequent sections provide a comprehensive overview of the three distinct categories of cavitation devices.

7.1.1.3 Effects

- Erosion refers to the process of considerable material erosion caused by the collapse of bubbles near surfaces.
- Cavitation can produce significant levels of noise and vibrations.
- *Efficiency degradation*: It might result in a decrease in the efficiency of pumps and turbines.

7.1.2 ACOUSTIC CAVITATION: DEFINITION

Acoustic cavitation is the phenomenon that occurs when a liquid is exposed to intense ultrasonic vibrations, resulting in the creation, enlargement, and sudden collapse of bubbles. This phenomenon is caused by the fluctuations in pressure of the sound waves. Ultrasound waves at a frequency of 16–100 MHz can induce acoustic cavitation conditions, resulting in pressure fluctuations in the liquid medium. Acoustic cavitation is being increasingly recognized as a very effective processing tool with broad applications in various fields of research and engineering. Acoustic waves induce pressure oscillations, leading to the creation and expansion of bubbles in this particular form of cavitation. When bubbles reach a certain size, they become unstable and collapse, resulting in a substantial release of energy in the form of mechanical, chemical, and thermal energy into the surrounding liquid. High-speed micro-jets, with velocities exceeding 150 m/s, are formed due to the fast fluctuations in the area of the vapor-liquid bubble interface. When a micro-jet forcefully collides with the opposing bubble wall, it generates a series of shock waves traveling at an approximate average speed of 2,000 m/s. Subsequently, more powerful shock waves are released. Ultimately, the gas and vapor contained within the bubble undergo compression, leading to the formation of localized regions with extremely high temperature (5000 K) and pressure (500 atm). These hot spots have the ability to induce the breakdown of chemical molecules into highly reactive radicals through a process known as sonolysis (Figure 7.2). Inside a bubble, the combination of high temperature and pressure creates the necessary energy to break the bonds of the gasified solutes and solvents in a way that splits the atoms evenly. This process generates radicals, which are highly reactive and can combine with each other to form new radicals and molecules.

These newly formed substances can then spread throughout the surrounding liquid, where they can react with the larger body of the liquid. Moreover, the ultrasonic field can significantly impact heat transfer methods, including convective and subcooled boiling heat transfer. Ultrasonic waves induce acoustic cavitation, resulting in a higher concentration of tiny bubbles on the surface compared to their absence. Upon the collapse of cavitation bubbles, a substantial amount of energy is liberated, leading to heightened turbulence and facilitating reaction rates through the generation of minute motions within the medium. The collapse of the bubble results in the formation of highly intense temperatures and pressures. Moreover, the number of unpaired atoms (free radicals) generated during the cavitation process can be controlled by manipulating specific parameters in the operation of a sonochemical reactor.

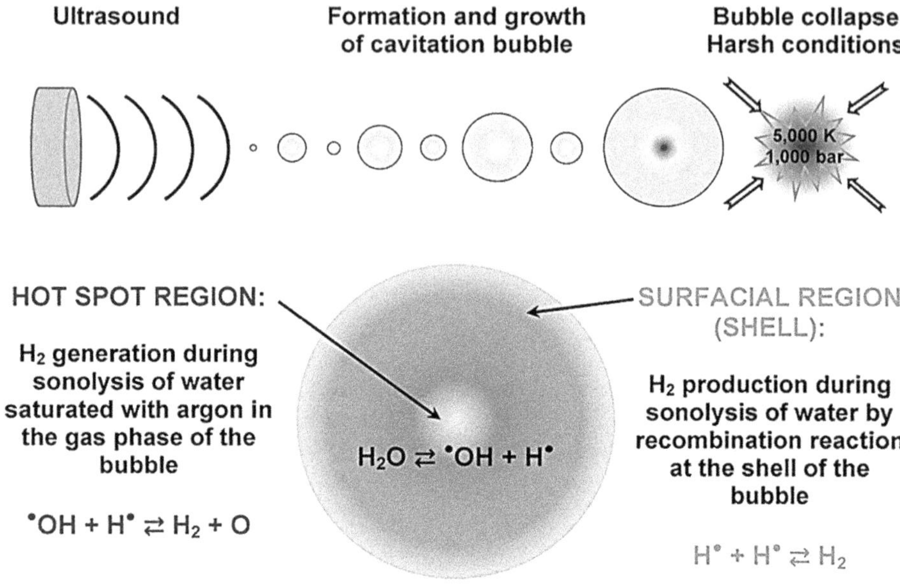

$$H_2O \rightleftarrows {}^{\bullet}OH + H^{\bullet}$$

$${}^{\bullet}OH + H^{\bullet} \rightleftarrows H_2 + O$$

$$H^{\bullet} + H^{\bullet} \rightleftarrows H_2$$

Hot spot inside collapsing acoustic bubble

FIGURE 7.2   Acoustic cavitation and sonolysis chemical reaction mechanism.[2]

7.1.2.1   Mechanism

- *Pressure fluctuations*: A liquid experiences alternating cycles of high and low pressure as a result of the propagation of ultrasonic vibrations.
- *Bubble formation*: During the phase of low pressure (rarefaction), tiny gas nuclei present in the liquid expand and develop into bubbles.
- *Bubble expansion*: These bubbles increase in size over consecutive cycles of reduced pressure.
- Bubble collapse refers to the process in which bubbles implode forcefully during the phase of high pressure, resulting in the generation of intense localized temperatures and pressures.

7.1.2.2   Effects

- Sonoluminescence refers to the emission of light resulting from the collapse of bubbles.
- Chemical reactions can be initiated by high temperatures and pressures, which are particularly beneficial in processes such as sonochemistry.
- Acoustic cavitation is employed in ultrasonic cleaning to eliminate impurities from surfaces.

7.1.3   Sonolysis

Sonolysis, often known as ultrasonic irradiation, is a prominent chemical procedure that involves the breakdown of chemical compounds into extremely reactive radicals.

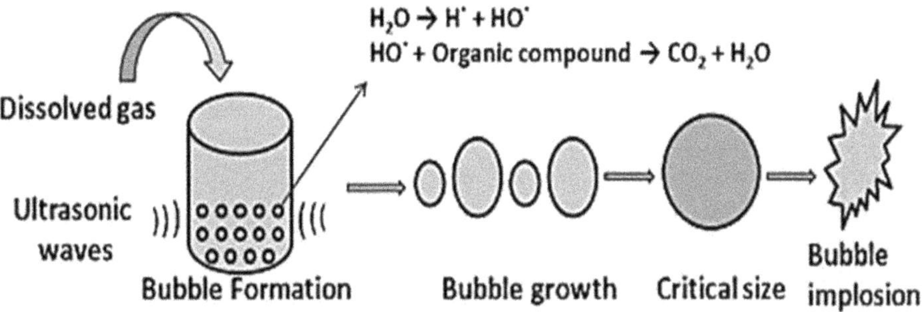

FIGURE 7.3   Sonolysis of waste water and degradation.[3]

Acoustic cavitation generates extreme circumstances, including high temperatures and pressures, which result in the creation of extremely reactive radicals. Below are several instances of chemical compounds that undergo sonolysis, resulting in the formation of active radicals (Figure 7.3). Additionally, the mechanisms and applications of these reactions are provided:

Examples

WATER (H_2O)

Mechanism:

- Ultrasonic waves cause the cavitation of water, leading to the formation of hydroxyl radicals (O$\dot{\text{H}}$) and hydrogen atoms ($\dot{\text{H}}$).

$$H_2O + \text{ultrasonication} \rightarrow O\dot{H} + \dot{H} \qquad (7.2)$$

Applications:

- *Advanced oxidation processes (AOPs)*: Hydroxyl radicals are highly reactive and are used to degrade pollutants in water treatment.
- *Sonochemistry*: These radicals initiate various chemical reactions, enhancing the rate and efficiency of processes like polymerization and organic synthesis.

HYDROGEN PEROXIDE (H_2O_2)

Mechanism:
- Sonolysis of hydrogen peroxide produces hydroxyl radicals.

$$H_2O_2 + \text{ultrasonication} \rightarrow 2O\dot{H} \qquad (7.3)$$

Applications:

- *Disinfection*: The hydroxyl radicals generated are potent disinfectants, capable of destroying bacteria and viruses.
- *Pollutant degradation*: Used in environmental remediation to break down complex organic pollutants.

7.2 INTRODUCTION TO HYDRODYNAMIC CAVITATION IN MICROCHANNELS

Microchannels are small and narrow channels via which fluids flow. These channels generally have dimensions ranging from tens to hundreds of micrometers (μm). Microchannels are frequently utilized in a wide range of applications because of their elevated surface area to volume ratio, which improves heat-and-mass transfer processes. Below is an elaborate elucidation of microchannels within the framework of hydrodynamic cavitation:

7.2.1 Microchannels

Microchannels are narrow and small-scale channels that are typically used for fluid flow. They are characterized by their small dimensions, which are typically in the range of micrometers to millimeters. Microchannels are commonly used in various applications, such as microfluidics, heat exchangers (Figure 7.4), and chemical reactors.

- Microchannels are small-scale channels specifically designed to facilitate the transportation of fluids. They possess the following distinguishing features:
- Microchannels typically have dimensions ranging from 10 to 500 μm in width and height.

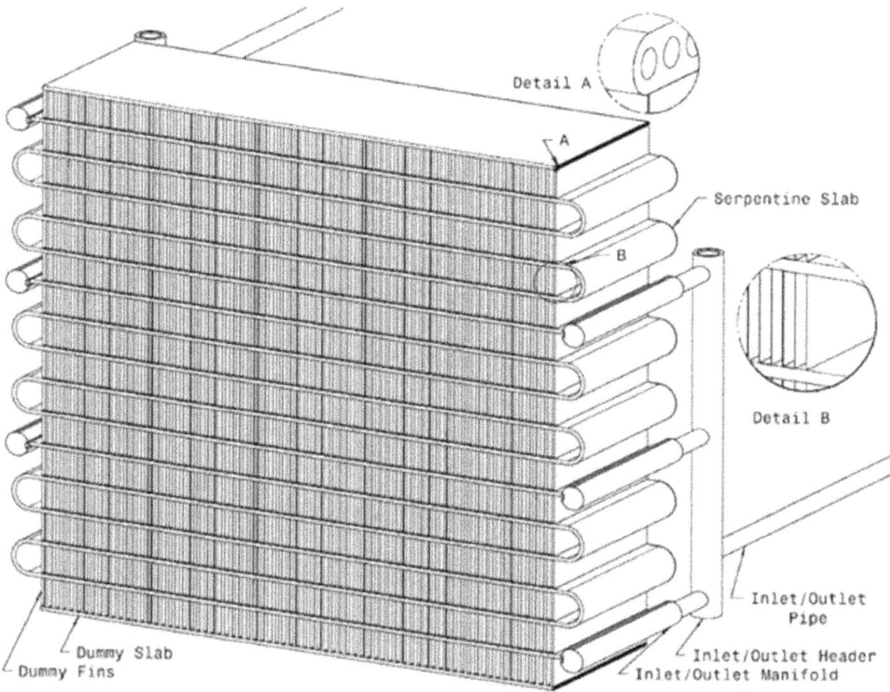

FIGURE 7.4 Microchannel heat exchanger.[4]

- The high surface area to volume ratio greatly enhances the effectiveness of heat-and-mass transfer processes.
- *Restricted flow*: The flow of fluid in microchannels is tightly limited, resulting in distinct fluid dynamics when compared to larger channels.

7.2.2 STUDY OF THE PHENOMENON OF HYDRODYNAMIC CAVITATION OCCURRING IN MICROCHANNELS

Hydrodynamic cavitation in microchannels entails the creation, enlargement, and implosion of vapor bubbles inside the fluid flow as a result of abrupt pressure fluctuations. This phenomenon can be purposely induced or unintentionally occur, depending on the fluid dynamics and channel design. The impact of hydrodynamic cavitation in microchannels is especially significant because of the restricted flow conditions and the intensified interaction between bubbles and channel walls.

7.2.2.1 The Mechanism of Cavitation in Microchannels

- *Pressure drop*: When a fluid passes through a microchannel, it may come across narrow sections or abrupt changes in direction, causing the fluid to speed up and the pressure to decrease in that specific area.
- Vapor bubble formation occurs when the pressure falls below the vapor pressure of the fluid.
- Bubble dynamics refer to the phenomenon where bubbles undergo growth, movement, and final collapse when they traverse regions of elevated pressure inside the channel.
- The bursting of these bubbles results in the release of substantial energy in the form of concentrated high temperatures and pressures.

7.2.3 FACTORS INFLUENCING CAVITATION IN MICROCHANNELS

Several variables impact the frequency and consequences of hydrodynamic cavitation in microchannels:

- The configuration and measurements of the microchannel, which include narrowing's and curves, impact the distribution of pressure and the dynamics of flow.
- Fluid properties such as viscosity, surface tension, and vapor pressure play a crucial role in determining the likelihood of cavitation.
- *Flow rate*: Increased flow rates might result in higher pressure drops and encourage the occurrence of cavitation.
- Surface roughness can act as nucleation sites for bubble production on the walls of the channel.

7.3 EFFECT OF HYDRODYNAMIC CAVITATION ON BOILING HEAT TRANSFER IN MICROCHANNELS

Hydrodynamic cavitation has a substantial effect on the transfer of heat during boiling in microchannels. This phenomenon is crucial for improving thermal management systems in different applications, including cooling microelectronics, medical equipment, and energy systems. To comprehend this effect, one must examine the interaction between fluid dynamics, phase transition phenomena, and thermal characteristics at the microscale level.

7.3.1 HEAT TRANSFER DURING BOILING IN MICROCHANNELS

Boiling heat transfer in microchannels occurs when a liquid undergoes a phase shift to vapor in small spaces with a high ratio of surface area to volume. This increases the rates at which heat is transferred compared to traditional systems on a larger scale. The mechanisms consist of nucleate boiling, which involves the formation of bubbles at specific places and their growth owing to heat transfer from the surface, and film boiling, which occurs when a continuous vapor layer develops, insulating the surface and decreasing the efficiency of heat transfer.

7.3.2 STUDY ON THE EFFECTS OF HYDRODYNAMIC CAVITATION ON BOILING HEAT TRANSFER

Hydrodynamic cavitation in microchannels significantly affects boiling heat transfer through various mechanisms:

- *Nucleation enhancement*: The presence of cavitation bubbles facilitates the formation of nucleation sites, which in turn initiate the process of boiling at reduced temperatures and consequently improve the rates of heat transfer.
- Bubble dynamics refers to the fast creation and collapse of cavitation bubbles, which can disturb thermal boundary layers and improve convective heat transfer.
- Pressure fluctuations caused by cavitation can affect the different stages of boiling, causing a transition between nucleate boiling and film boiling, and vice versa.
- *Thermal conductivity*: The fluid's ability to conduct heat can be temporarily enhanced by the localized high temperatures resulting from bubble collapse.

7.3.3 MECHANISMS OF INFLUENCE

7.3.3.1 Facilitation of Nucleation

Hydrodynamic cavitation enhances the creation of vapor bubbles by offering supplementary locations for nucleation. The presence of microchannels limits the number of natural nucleation sites due to the constrained shape. The presence of cavitation bubbles greatly enhances the accessible surface area for nucleation, resulting in an earlier initiation of boiling and increased rates of heat transfer.

Example

In microelectronics cooling systems, the occurrence of boiling at an early stage, facilitated by cavitation, assists in maintaining lower operating temperatures. This, in turn, enhances the reliability and performance of electronic components.

7.3.3.2 Study of the Behavior of Bubbles and the Transfer of Heat

The phenomena of cavitation bubbles, encompassing their expansion, merging, and implosion, have a pivotal impact on augmenting heat transport. The bursting of bubbles creates micro-jets and shock waves, which disturb the thermal boundary layer and enhance turbulent mixing. This increases the convective heat transfer coefficient, resulting in enhanced total heat transfer performance.

Example

Microchannel heat sinks utilized in high-performance computing systems benefit from improved convective heat transfer resulting from the dynamics of cavitation bubbles. This better heat dissipation enables higher processing power without the risk of overheating.

7.3.3.3 Variations in Pressure and the Transition between Boiling Regimes

Cavitation can generate fast pressure variations that lead to transitions between several boiling regimes. For example, cavitation can disrupt film boiling by rupturing the vapor film and transitioning to nucleate boiling, which is a more effective method of transferring heat. As illustrated in Figure 7.5, the exit pressure fluctuated throughout a range that corresponded to three distinct flow patterns: single-phase flow transition to cavitating flow, cavitating flow pattern, and bubbly flow pattern. In addition, inside the fully developed cavitation flow, there were a number of different flow morphologies present along the channel. The flow patterns that are being discussed here can be broken down into five distinct categories: (1) liquid single-phase flow, (2) liquid jet flow, (3) transition zone flow, (4) annular and wavy-annular flow, and (5) bubbly flow.

Example

Microchannel-based refrigeration systems can optimize the cooling cycle and improve the coefficient of performance (COP) by managing the boiling regime through cavitation.

7.3.3.4 Increase in Thermal Conductivity

The localized elevation of temperatures and pressures resulting from the collapse of cavitation bubbles might temporarily enhance the thermal conductivity of the fluid. The impact of this phenomenon is especially notable in microchannels, as the limited amount of fluid enables quick thermal equalization.

Example

Microreactors utilized in chemical synthesis benefit from better thermal conductivity resulting from cavitation, which in turn boosts heat transfer rates, resulting in accelerated reaction kinetics and increased product yields.

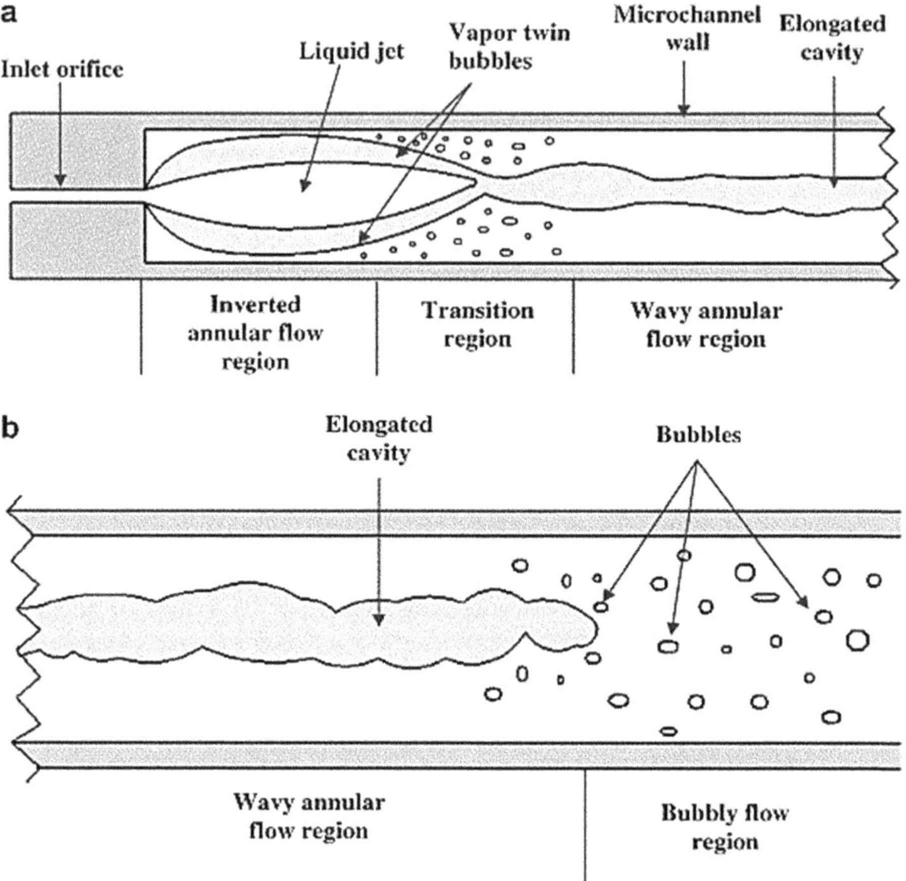

FIGURE 7.5 (a) Bubble growth and (b) bubble collapse due to hydrodynamic cavitation in a hydrodynamic orifice reactor.[5]

7.3.4 EXPERIMENTAL AND NUMERICAL STUDIES

7.3.4.1 Investigations Involving Both Experimentation and Numerical Analysis

Several empirical and computational investigations have been carried out to comprehend the impact of hydrodynamic cavitation on the process of heat transfer during boiling in microchannels. These investigations entail altering the channel's geometry, fluid characteristics, and operating circumstances to examine the resulting heat transfer efficiency.

7.3.4.2 Experimental Studies

Case Study 1: Microchannel Geometry and Cavitation

A research study was conducted to examine how the shape of microchannels affects the process of cavitation and boiling heat transfer. Various microchannel configurations, such as straight, converging-diverging, and serpentine channels, were subjected to testing. The study discovered that the utilization of converging-diverging channels resulted in an augmentation of cavitation, which in turn led to greater nucleate boiling and increased rates of heat transfer as compared to straight channels.

Case Study 2: Investigation of Fluid Properties and Cavitation

A separate investigation examined the impact of fluid characteristics, such as viscosity and surface tension, on the occurrence of cavitation and the transfer of heat through boiling in microchannels. Fluids possessing reduced viscosity and surface tension displayed more prominent cavitation phenomena, leading to improved heat transfer during boiling.

7.3.4.3 Numerical Studies

7.3.4.3.1 Simulation 1: Cavitation Bubble Dynamics

Computational simulations were performed to examine the dynamics of cavitation bubbles and their influence on heat transfer through boiling in microchannels. The simulations employed computational fluid dynamics (CFD) models to accurately depict the development, implosion, and interaction of cavitation bubbles. The findings indicated that the existence of cavitation bubbles substantially improved convective heat transfer by disturbing the thermal boundary layer.

7.3.4.3.2 Simulation 2: Transitions in Boiling Regimes

Numerical research examined the impact of pressure variations caused by cavitation on the transition between different boiling regimes in microchannels. The simulation findings demonstrated that cavitation has the potential to cause fast shifts between nucleate boiling and film boiling, hence improving the overall efficiency of heat transfer by keeping the system in the most favorable boiling state.

7.3.5 PRACTICAL APPLICATIONS

7.3.5.1 Thermal Management of Microelectronics

Efficient thermal management is essential in microelectronics to provide reliable performance and prevent overheating. Hydrodynamic cavitation occurring in microchannel heat sinks improves the process of boiling heat transfer, hence enabling the dissipation of heat at faster rates. This is especially crucial for high-power electrical components, such as CPU (Central Processing Unit) and a GPU (Graphics Processing Unit), when conventional cooling methods are inadequate.

Example

An innovative microchannel heat sink design was created to efficiently cool high-power microprocessors by adding elements that promote cavitation. Cavitation-induced enhancement in boiling heat transfer resulted in a substantial improvement in the ability to dissipate heat, hence enabling faster processing speeds without experiencing thermal throttling.

7.3.5.2　Medical Devices

Precise heat control is crucial in medical equipment, particularly for applications such as tissue ablation and hyperthermia treatment. Hydrodynamic cavitation occurring in microchannels can improve the efficiency of localized boiling heat transfer, leading to improved thermal management in certain applications.

Example

A thermal therapy device for cancer treatment, which is built on microchannels, utilizes cavitation to improve the process of boiling heat transfer. The enhanced heat transfer facilitated accurate temperature regulation, hence improving the efficacy of the treatment while reducing harm to adjacent healthy tissues.

7.3.5.3　Power Systems

Enhanced boiling heat transfer in energy systems, such as microchannel heat exchangers used for power production and refrigeration, leads to improved system efficiency and performance. Hydrodynamic cavitation can significantly enhance the efficiency of phase transition processes, making it crucial for improving these systems.

Example

A microchannel heat exchanger was specifically developed for a tiny power generation system to include elements that promote cavitation. Cavitation-induced enhancement in boiling heat transfer resulted in an overall improvement in the thermal efficiency of the system, enabling higher power generation while consuming less fuel.

7.3.6　CHALLENGES AND FUTURE DIRECTIONS

Although hydrodynamic cavitation presents notable advantages for enhancing boiling heat transfer in microchannels, it also poses obstacles and opportunities for future investigation:

7.3.6.1 Challenges

- Material erosion can occur in microchannels due to the forceful collapse of cavitation bubbles, which has the potential to decrease the system's lifespan.
- Controlling cavitation with precision in order to enhance heat transmission while avoiding damage is a difficult task that necessitates the use of sophisticated design and control technologies.
- *Scalability*: Scaling up microchannel systems with cavitation for industrial applications requires overcoming engineering and manufacturing challenges.

7.3.6.2 Future Directions

- Advanced materials research focuses on the development of erosion-resistant materials and coatings to minimize the negative impacts of cavitation-induced erosion.
- Implementing intelligent control systems to dynamically modify operating conditions and optimize cavitation for improved heat transfer.
- Hybrid cooling systems involve the integration of hydrodynamic cavitation with other cooling approaches, such as jet impingement or phase change materials, in order to augment the efficiency of heat transmission.
- Advancing multiphase flow modeling tools to enhance the accuracy and efficiency of predicting and optimizing cavitation and boiling heat transfer in microchannels.

7.4 INFLUENCE OF ACOUSTIC CAVITATION ON BOILING HEAT TRANSFER

Boiling heat transfer is extensively employed in energy resource, electricity, aviation, and space travel operations due to its exceptional heat transfer capacity. Nevertheless, the escalating heat fluxes in advanced equipment necessitate more effective and dependable cooling techniques. In order to effectively regulate the process of heat transfer during boiling, it is imperative that we possess a comprehensive understanding of the phenomenon of boiling. Despite the abundance of experimental data on heat fluxes and wall superheats, and the existence of many experimental expressions that relate to the nucleation and growth of bubbles, no efficient strategies have been discovered to regulate boiling heat transfer.

Cavitation is the physical phenomenon in liquids that occurs when the pressure is reduced at a relatively constant liquid temperature, resulting in the rupture of the liquid. Acoustic cavitation is the process in which tiny bubbles are formed, expand, and collapse within a liquid as a result of sonic stimulation. Wong and Chon[6] in 1969 conducted an experiment to determine the impact of ultrasonic vibrations on boiling heat transfer. They found that this impact became insignificant when the boiling process reached its fully developed nucleate boiling stage. Later in 2004, Zhou and Liu[7] conducted a comprehensive study on the impact of sound source intensity, distance,

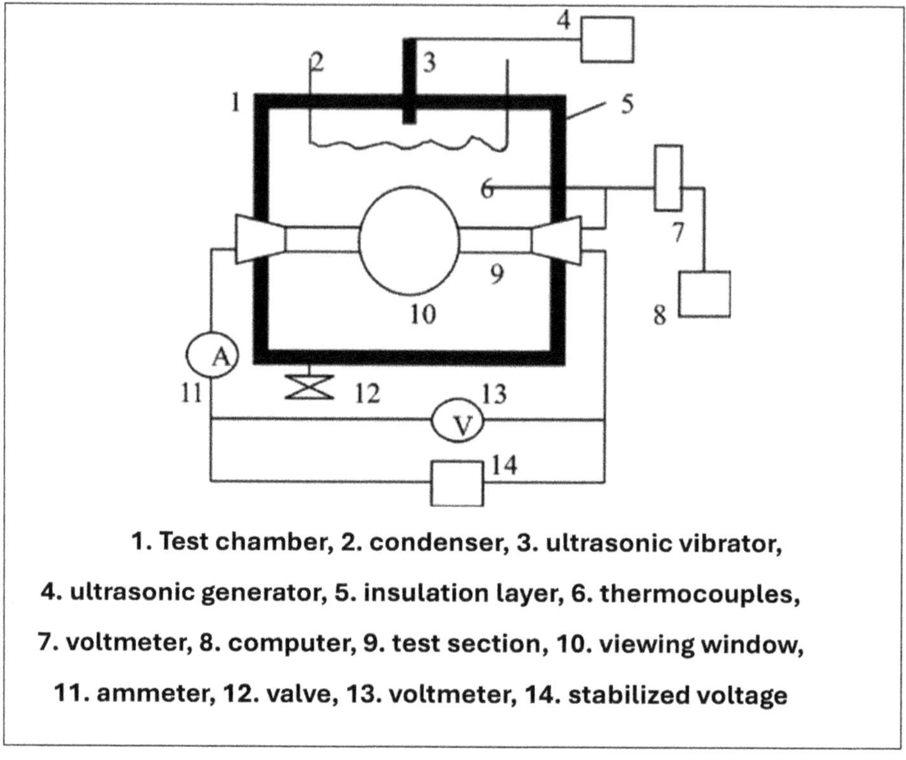

FIGURE 7.6   Experimental section for the acoustic cavitation-induced boiling.[8]

vibration location, and fluid subcooling on boiling heat transfer on a horizontal circular copper tube. Studies have shown that the boiling heat transfer can be either enhanced or reduced depending on the intensity of the sound source. However, they provided solely a mechanical explanation for the effects. Additionally, researchers have investigated the influence of acoustical field parameters and $CaCO_3$ nanoparticles with a size less than 100 nm, in the form of nanofluids, on the process of boiling heat transfer in a pool of acetone. The test part was depicted in Figure 7.6. It was confirmed that the acoustic cavitation improved the dispersion of nanoparticles, resulting in the formation of a stable nanofluid solution within 30 minutes of operation.

7.4.1 Types of Boiling: Pool and Flow Boiling

Boiling can be categorized into two primary forms: pool boiling and flow boiling as shown in Figure 7.7. These classifications are determined by whether the liquid remains still or moves across a heated surface. Each category can be further subdivided according to distinct traits and patterns of boiling behavior. Boiling can be classified into two main categories: pool boiling and flow boiling. Below is a comprehensive explanation of these two types.[9]

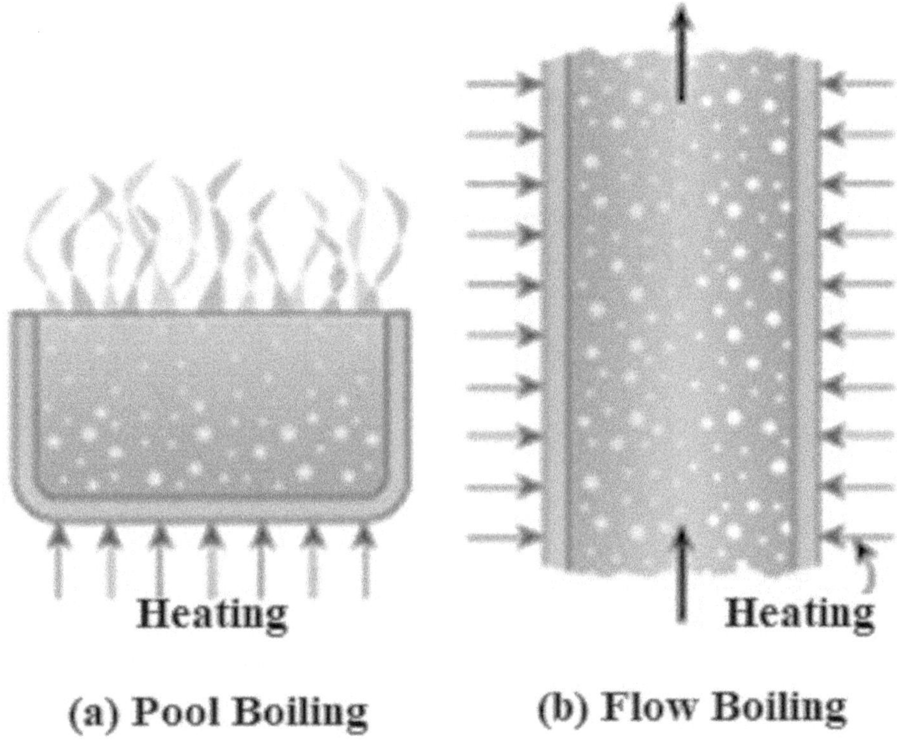

FIGURE 7.7   Type of boiling based on bulk fluid motion.

7.4.1.1   Pool Boiling

Pool boiling refers to the phenomenon when a liquid remains still and is heated by a surface that is submerged in the liquid. The main categorization in pool boiling is determined by the heat flux and temperature disparities, resulting in distinct boiling regimes as illustrated in Figure 7.8.

7.4.1.1.1   Natural Convection Boiling

Definition: This phenomenon takes place when there is a low amount of heat being transferred and the temperature of the liquid is lower than its boiling point. Heat is transmitted through natural convection currents, without the creation of large bubbles.

   Characteristics: There are no bubbles present, and the primary method of heat transport is through convection.

7.4.1.1.2   Nucleate Boiling

Nucleate boiling refers to the process of boiling that occurs when little bubbles of vapor form on a heated surface.

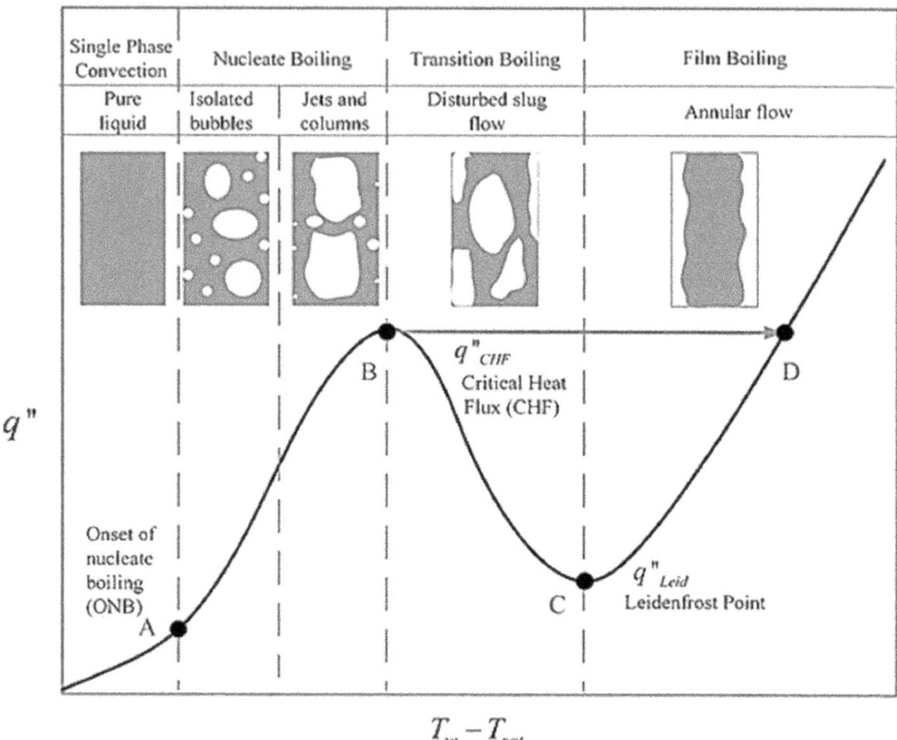

FIGURE 7.8   Typical pool boiling curve.[10]

Definition: When the rate of heat transfer rises, bubbles begin to develop at certain locations on the heated surface called nucleation sites.

Characteristics: High heat transfer rates due to the latent heat of vaporization. The processes of bubble creation, development, and separation are the main factors that control heat transport.

7.4.1.1.3   Transition Boiling

Transition boiling refers to the process of a liquid changing into a vapor due to an increase in temperature or pressure.

Definition: A transitional state between nucleate boiling and film boiling, distinguished by volatile boiling conditions.

Characterization: The process of bubbles merging together to form vapor patches, resulting in varying rates of heat transfer.[11]

7.4.1.1.4   Film Boiling

Definition: When subjected to extremely high levels of heat, a heated surface develops a continuous layer of vapor that acts as insulation, preventing direct contact with the liquid.

Characteristics: The heat transmission rates are reduced because of the insulating action of the vapor coating. Formation of a vapor film and establishment of a steady heat transmission regime.[12]

7.4.1.2 Flow Boiling

Flow boiling refers to the process of boiling a liquid while it is flowing through a conduit or channel. Flow boiling is the phenomenon that occurs when a liquid moves across a heated surface, and boiling is caused by the combined influence of heat flux and fluid motion. Flow boiling can be categorized according to the fluid's state (either subcooled or saturated) and the flow patterns it exhibits.

7.4.1.2.1 Subcooled Flow Boiling

Subcooled flow boiling refers to the process of boiling a liquid at a temperature below its saturation point while it is flowing through a system. The liquid is at a temperature lower than its saturation temperature, causing bubbles to form and collapse inside the fluid.

Characteristics: Combined effects of convection and boiling. Bubbles are generated, expand, and collapse within the fluid in motion.

7.4.1.2.2 Saturated Flow Boiling

Saturated flow boiling refers to the process of boiling a liquid at its saturation temperature, where the liquid and vapor phases coexist.[13]

Definition: The temperature of the liquid reaches the point at which it can no longer hold any more heat, causing bubbles to appear on the surface that is being heated.

Characteristics: High heat transfer rates due to continuous bubble formation and removal by the flowing liquid.

7.4.1.3 Stable Boiling Condition

7.4.1.3.1 Annular Flow Boiling

It refers to the flow pattern in which a liquid or gas flows in a ring-shaped channel, with a central core of one phase surrounded by a thin film of another phase. At elevated heat fluxes, a slender layer of liquid forms on the heated surface while vapor flows in the center.

Characteristics: Exceptionally efficient heat transfer. There is a possibility that the liquid film may dry out, which can result in critical heat flux circumstances.

7.4.1.4 Key Differences between Pool Boiling and Flow Boiling

7.4.1.4.1 Pool Boiling

- *Liquid state*: Stationary liquid.
- *Heat transfer mechanism*: Dominated by bubble dynamics and natural convection.
- *Boiling regimes*: Natural convection, nucleate boiling, transition boiling, and film boiling.
- *Applications*: Heat exchangers, nuclear reactors, and immersion cooling.

7.4.1.4.2　Flow Boiling

- *Liquid state*: Flowing liquid.
- *Heat transfer mechanism*: Combined effects of convection and boiling.
- *Boiling regimes*: Subcooled flow boiling, saturated flow boiling, and annular flow boiling.
- *Applications*: Heat exchangers, refrigeration and air conditioning, and power generation.

7.4.1.5　Influence of Nanofluids on Pool and Flow Boiling

7.4.1.5.1　Pool Boiling with Nanofluids

- *Enhanced thermal conductivity*: Nanoparticles increase the thermal conductivity of the base fluid, improving heat transfer rates.
- *Increased nucleation sites*: Nanoparticles provide additional nucleation sites, leading to more efficient bubble formation and growth.
- *Improved critical heat flux (CHF)*: Higher CHF due to enhanced thermal properties and better heat distribution.[14]

7.4.1.5.2　Flow Boiling with Nanofluids

- *Increased heat transfer coefficient*: Nanoparticles enhance convective and boiling heat transfer, leading to higher heat transfer rates.
- *Delayed CHF*: Improved thermal properties and higher flow rates delay the onset of critical heat flux conditions.
- *Stable boiling regimes*: Nanoparticles contribute to stable boiling behavior, reducing temperature fluctuations and improving efficiency.

7.5　SUMMARY

- Hydrodynamic cavitation provides notable advantages for the process of boiling heat transfer in microchannels, resulting in improved heat dissipation and enhanced thermal management across a range of applications.
- Nevertheless, the general implementation of this technology is hindered by hurdles such as material erosion, noise, control difficulties, and scaling issues. Promising solutions to these restrictions lie in future research and technical breakthroughs, such as the creation of materials that are resistant to erosion, the implementation of intelligent control systems, the integration of nanofluids, the use of advanced computational models, and the adoption of hybrid cooling systems.
- Boiling is classified into two categories: pool boiling and flow boiling.
- Pool boiling refers to the process when the liquid is stationary over a heated surface, whereas flow boiling occurs when the liquid is in motion over the heated surface.
- Each category exhibits distinct boiling patterns distinguished by varying heat transmission processes and behaviors.
- Nanofluids improve thermal conductivity, increase nucleation sites, and enhance overall heat transfer performance in both pool and flow boiling processes.

REFERENCES

1. Hamidi, R., Damizia, M., De Filippis, P., Patrizi, D., Verdone, N., Vilardi, G., & de Caprariis, B. (2023). Recent developments and future outlooks of hydrodynamic cavitation as an intensification technology for renewable biofuels production. *Journal of Environmental Chemical Engineering*, 11(5).
2. Kerboua, K., & Hamdaoui, O. (2019). Energetic challenges and sonochemistry: A new alternative for hydrogen production? *Current Opinion in Green and Sustainable Chemistry*, 18, 84–89.
3. Katiyar, J., Bargole, S., George, S., Bhoi, R., & Saharan, V. K. (2021). Advanced technologies for wastewater treatment: New trends. In *Handbook of Nanomaterials for Wastewater Treatment* (pp. 85–133). Amsterdam: Elsevier. https://doi.org/10.1016/B978-0-12-821496-1.00009-X.
4. Siddiqui, F. (2011). A study of cross-flow air heating via a multiport serpentine microchannel heat exchanger. MSc. dissertation, Dept. Mechanical Automotive and Materials Eng., University of Windsor, Ontario, Canada.
5. Schneider, B., Koşar, A., & Peles, Y. (2007). Hydrodynamic cavitation and boiling in refrigerant (R-123) flow inside microchannels. *International Journal of Heat and Mass Transfer*, 50(13–14), 2838–2854.
6. Wong, S. W., & Chon, W. Y. (1969). Effects of ultrasonic vibrations on heat transfer to liquids by natural convection and by boiling. *AIChE Journal*, 15(2), 281–288. https://doi.org/10.1002/aic.690150229
7. Zhou, D.W., & Liu, D. Y. (2002). Boiling heat transfer in an acoustic cavitation field. *Chinese Journal of Chemical Engineering,* 10, 100–105.
8. Zhou, D., Liu, D., Hu, X., & Ma, C. (2002). Effect of acoustic cavitation on boiling heat transfer. *Experimental Thermal and Fluid Science*, 26(8), 931–938. https://doi.org/10.1016/s0894-1777(02)00201-7
9. Thakur, P., Sonawane, S., Bhaisare, S., & Pandey, N. (2021). Enhancement of pool boiling performance using SWCNT based nanofluids: A sustainable method for the wastewater heat recovery. *Journal of Indian Association for Environmental Management (JIAEM)*, 41(4), 7–18.
10. Thakur, P. P., Khapane, T. S., & Sonawane, S. S. (2021). Comparative performance evaluation of fly ash-based hybrid nanofluids in microchannel-based direct absorption solar collector. *Journal of Thermal Analysis and Calorimetry*, 143, 1713–1726.
11. Kumar, N., Urkude, N., Sonawane, S. S., & Sonawane, S. H. (2018). Experimental study on pool boiling and Critical Heat Flux enhancement of metal oxides based nanofluid. *International Communications in Heat and Mass Transfer*, 96, 37–42.
12. Hu, H., Xu, C., Zhao, Y. et al. (2017). Boiling and quenching heat transfer advancement by nanoscale surface modification. *Science Report, 7,* 6117. https://doi.org/10.1038/s41598-017-06050-0
13. Kumar, N., Urkude, N., Sonawane, S. S., & Sonawane, S. H. (2018). Experimental study on pool boiling and Critical Heat Flux enhancement of metal oxides based nanofluid. *International Communications in Heat and Mass Transfer*, 96, 37–42.
14. Kumar, N., Sonawane, S. S., & Sonawane, S. H. (2018). Experimental study of thermal conductivity, heat transfer and friction factor of Al_2O_3 based nanofluid. *International Communications in Heat and Mass Transfer*, 90, 1–10.

8 Ultrasound-Assisted Nanofluid Flooding to Enhance Heavy Oil Recovery

8.1 INTRODUCTION TO OIL RECOVERY

The recovery process of petroleum reservoir consists of three distinct phases:

- primary,
- secondary, and
- tertiary.

8.1.1 PRIMARY RECOVERY

During the primary recovery phase, oil is extracted as a result of the natural energy present in the reservoir, such as compaction drive, solution gas drive, water drive, gas cap drive, and gravity drive. Primary recovery is dependent on the inherent pressure of the reservoir to facilitate the upward movement of oil to the surface. This phase typically retrieves a *mere 10%–30% of the initial oil in place* and encompasses the subsequent mechanisms:

8.1.1.1 Natural Drive Mechanisms

- *Solution gas drive*: When oil is extracted from underground, the decrease in pressure causes gases that were dissolved in the oil to separate, forming a gas cap. This gas cap then aids in pushing the oil toward the wellbore.
- *Gas cap drive*: It is a phenomenon where a layer of natural gas located above the oil column applies pressure, causing the oil to move downward and toward the production wells.
- *Water drive*: The presence of water below the oil zone creates a force that pulls the oil upward due to the upward pressure exerted by the water.
- *Gravity drainage*: The force of gravity facilitates the downward movement of oil within the reservoir, directing it toward the production wells.

8.1.1.2 Artificial Lift Techniques

When the natural pressure is inadequate to propel oil to the surface, artificial lift methods such as rod pumping (beam pumping) or gas lift (injecting gas into the well to decrease the density of the oil) are used to increase production.

DOI: 10.1201/9781003594949-8

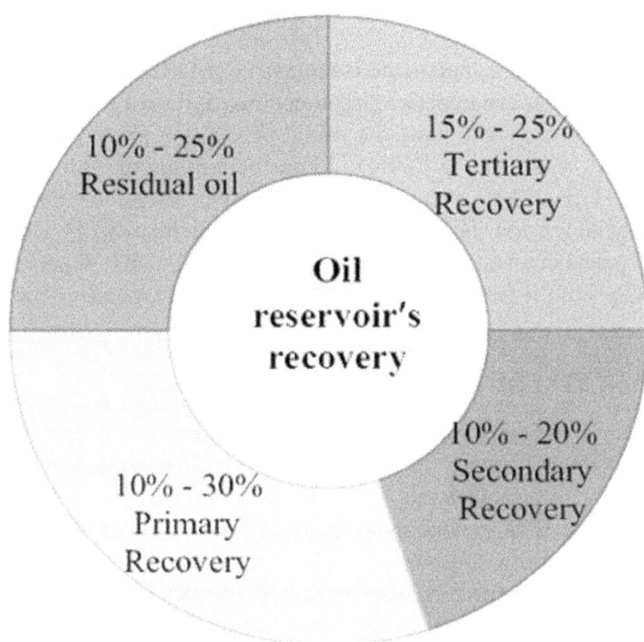

FIGURE 8.1 Available primary, secondary, and tertiary techniques for the extraction of oil.[1]

8.1.2 Secondary Recovery

When the primary methods of recovery become ineffective, secondary recovery procedures are employed. These techniques are designed to enhance the pressure within the reservoir and enhance the movement of oil toward the wellbore, typically resulting in the recovery of an extra 20%–40% of the original oil in place. The implementation of these strategies results in an additional recovery factor of 10%–20% (Figure 8.1).

8.1.2.1 Recovery Mechanisms

8.1.2.1.1 Water Flooding

Process: Reservoirs are infused with water using injection wells. This action raises the pressure in the reservoir, facilitating the movement of oil toward the production wells.

Considerations: The success of water flooding depends on the geology of the reservoir, the viscosity of the oil, and the properties of the water and rock.

8.1.2.1.2 Gas Injection

Process: Natural gas, nitrogen, or carbon dioxide (CO_2) is injected into the reservoir to maintain pressure and displace oil toward the production wells.

Considerations: Gas injection can be employed to restore pressure in reservoirs that have experienced depletion as a result of primary recovery.

8.1.3 Tertiary Recovery

Tertiary methods, known as enhanced oil recovery (EOR) methods, are mostly used in mature oil fields that are experiencing a decline in output after primary and secondary methods, or soon following the initial production phase. "EOR" procedures involve the injection of gases or fluids to move the remaining oil trapped in the rock of the reservoir. This is done because of the strong viscosity and capillary forces and the high interfacial tension (IFT) between the fluid and the rock. The implementation of these approaches can result in a recovery rise of 15%–25%. Figure 8.1 illustrates the various stages of oil reservoir recovery and their corresponding recovery rates.

8.2 ENHANCED OIL RECOVERY TECHNIQUES

EOR refers to a collection of methods employed to augment the quantity of crude oil that may be collected from an oil field. Once the primary and secondary recovery techniques have been fully utilized, a substantial amount of the oil remains confined within the reservoir. EOR procedures are designed to extract the remaining oil that is left after primary and secondary recovery methods have been used. This residual oil usually accounts for approximately 30%–60% of the initial oil in the reservoir.

EOR approaches can be classified into two categories based on their distinct operational principles:

a. Techniques that enhance the efficacy of fluid injection in displacing oil at a microscopic level.
b. Techniques that enhance the efficiency of fluid injection to improve the volumetric sweep of a reservoir.

8.2.1 Enhancing the Oil Displacement Oil at a Microscopic Level

As per Equation (8.1), the effectiveness of microscopic oil displacement is defined as the ratio of the amount of oil that is produced to the amount of oil that is in touch with the fluid that is displacing it. The displacement efficiency is determined by factors such as IFT, wettability, capillary pressure (P_C-Equation 8.2), and relative permeability. The improvement in oil microscopic displacement efficiency can be done by reducing the oil viscosity by heating means or by reducing capillary force and IFT through the application of chemical treatments, such as injecting surfactants, alkali, or hydrocarbon gases like CO_2.

$$\text{Effectiveness of oil displacement} = \frac{\text{Amount of oil produced (or) recovered}}{\text{Oil that is in contact with displacing fluid}}, \quad (8.1)$$

$$\text{Capillary pressure } (P_C) = \gamma C = \frac{2\gamma \cos\theta}{R} \quad (8.2)$$

where γ is the interfacial tension (IFT), C is the curvature of the interface, θ is the contact angle, R is the pore diameter.

8.2.1.1 Important Definitions

8.2.1.1.1 Capillary Pressure

Capillary pressure (P_C) refers to the differential in pressure between two immiscible fluids at the interface within the pores of a reservoir rock. The parameter plays a critical role in regulating how fluids (such as oil, water, and gas) are distributed and move within the porous media of the reservoir. Capillary pressure exerts a notable impact on the saturation levels of these fluids and is a crucial factor in primary, secondary, and tertiary oil recovery techniques.

8.2.1.1.2 Interfacial Tension (IFT)

IFT, represented by the symbol γ, refers to the force per unit length acting at the interface between two immiscible phases.

Interfacial tension, often known as IFT, refers to the force exerted per unit length at the boundary between two fluids that do not mix, such as oil and water. The phenomenon occurs as a result of variations in the cohesive forces inside each fluid in comparison to the adhesive forces between the fluids. Regarding the process of oil recovery, managing interfacial tension (IFT) is crucial at every stage as discussed below,

- *Primary recovery*: A high IFT can hinder the passage of oil, resulting in a substantial amount of oil being trapped in the pores.
- *Secondary recovery*: The implementation of water flooding techniques can be enhanced by reducing IFT, which improves the efficiency of oil displacement.
- *Tertiary recovery techniques*: Like surfactant flooding, these techniques are employed to decrease IFT in order to facilitate the movement of oil and improve rates of recovery. Reducing the IFT diminishes the capillary forces that retain the oil, making it easier to displace and allow for smoother flow toward the producing wells.

8.2.1.1.3 Curvature of the Interface (C)

The curvature of the interface (C) characterizes the form of the meniscus that is created at the boundary between two fluids that do not mix, within the pore space. This pertains to the arrangement of pores and the way fluids interact with them.

- *Primary recovery*: The curvature of the reservoir is affected by the inherent circumstances, which in turn affects the distribution and movement of oil within the rock.
- *Secondary recovery*: The effectiveness of water in displacing oil from the pores during water flooding is determined by the curvature of the water-oil contact.
- *Tertiary recovery*: It involves the use of advanced techniques such as thermal recovery to modify the characteristics of the fluid and the curvature of the interface, resulting in improved displacement of oil. Heating the oil decreases its viscosity, altering the curvature and facilitating the flow.

8.2.1.1.4   The Contact Angle (Θ)

The contact angle (Θ) is the angle that is generated at the point of contact between the interface of a fluid and the solid surface of the rock in a reservoir. The statement refers to the measure of how easily a liquid can spread or adhere to the surface of a rock.

- *Primary recovery*: The contact angle has a crucial role in determining whether the rock is water-wet or oil-wet, which has a substantial impact on the distribution of oil and the efficiency of recovery. A water-wet rock refers to a rock surface where water has a greater affinity and sticks to it more readily, whereas an oil-wet rock indicates that oil has a stronger adhesion to the surface.
- *Secondary recovery*: Water-wet circumstances are generally more advantageous for water flooding, since the injected water can more effectively displace the oil. On the other hand, when conditions are oil-wet, it can impede the circulation of water and the displacement of oil.
- *Tertiary recovery*: It involves the use of techniques to adjust wettability, such as injecting surfactants, in order to change the contact angle to a more favorable state, typically making it more water-wet. This helps improve the displacement of oil and increases the amount of oil that can be recovered.

8.2.1.1.5   Pore Diameter (R)

The pore diameter (R) refers to the dimensions of the pores present in the reservoir rock. These pores determine the pathways and dispersion of fluids.

- *Primary recovery*: The early recovery rates and the amount of oil that remains trapped due to capillary forces are influenced by the distribution of pore sizes.
- *Secondary recovery*: The presence of smaller pore sizes might impede the passage of fluids and decrease the effectiveness of water flooding. Enlarged pores allow for better movement but can also result in the formation of channels and inefficient sweeping of the reservoir.
- *Tertiary recovery techniques*: Like polymer flooding, these techniques are employed to modify the viscosity of the injected fluids in order to more effectively align with the distribution of pore sizes. This adjustment enhances the effectiveness of fluid displacement and increases the amount of oil that can be recovered.

8.2.2   Enhancing the Volumetric Sweep Efficiency of a Reservoir

The volumetric sweep efficiency is determined by multiplying the vertical and areal sweep efficiency, which represent the portions of the reservoir area and thickness that are effectively swept by the displacing fluid. The efficiency of the process relies on several factors, including the heterogeneity of the reservoir, the mobility ratio between the injected fluid and oil, the pattern of injection and production wells, and the features of the reservoir rock. The macroscopic efficiency of oil displacement can be enhanced by reducing the mobility of the injected fluid, such as by injecting a polymer aqueous solution, or by increasing the mobility of the oil, such as using heating techniques.

8.2.2.1 Volumetric Sweep Efficiency (E_V)

Volumetric sweep efficiency (E_V) is a vital measure in EOR that evaluates how well an injected fluid, such as water, gas, or chemicals, displaces oil over the whole reservoir. It quantifies the proportion of the reservoir's pore volume that the injected fluid has effectively moved through. According to Equation (8.3), the overall volumetric sweep efficiency is determined by multiplying the areal sweep efficiency (E_A) and the vertical sweep efficiency (E_I).

$$\text{Overall volumetric sweep efficiency } \left(E_V\right) = E_A E_I \tag{8.3}$$

8.2.2.2 Areal Sweep Efficiency (E_A)

Areal sweep efficiency is a measure of the proportion of the reservoir's horizontal area that has been covered by the injected fluid. Reservoir heterogeneity, fluid characteristics, well spacing, and injection patterns are all factors that influence E_A.

8.2.2.2.1 *Estimation of Areal Sweep Efficiency*

- **Analytical Models**

    There are several ways available to estimate areal sweep efficiency:

    Analytical models employ mathematical formulas to forecast the areal sweep efficiency, taking into account fluid motion and reservoir properties.
- **Dykstra-Parsons Model**

    Dykstra-Parsons model incorporates reservoir heterogeneity by utilizing the coefficient of variation (V), which quantifies the degree of variability in permeability across the reservoir.

The Craig-Geffen-Morse method utilizes the mobility ratio (M) as a means to evaluate the areal sweep efficiency.

$$E_A = \frac{1}{1 + \ln\left(M\right)} \tag{8.4}$$

where

$$\text{Mobility of the fluid } \left(\lambda\right) = \frac{\text{Permeability}}{\text{Viscosity}} \tag{8.5}$$

$$\text{Mobility ratio } \left(M\right) = \frac{\text{Mobility of the displacing}}{\text{Mobility of the displaced}} \tag{8.6}$$

- **Simulation Models**

  Numerical reservoir simulation methods utilize geological, petrophysical, and fluid features to accurately replicate the behavior of injected fluids and forecast the efficiency of fluid displacement over a given area.

8.2.2.3 Vertical Sweep Efficiency (E_I)

Vertical sweep efficiency quantifies the proportion of the reservoir's vertical cross-section that has been covered by the injected fluid. Reservoir layering, changes in vertical permeability, and gravity segregation are all factors that influence E_I.

8.2.2.3.1 Vertical Sweep Efficiency Calculation

Vertical sweep efficiency can be approximated by utilizing the subsequent techniques:

Analytical Models

- Analytical models employ mathematical equations to forecast vertical sweep efficiency by considering fluid densities and injection rates.
- The Stiles method incorporates the effects of gravity segregation and fluid flow patterns in the reservoir layers.
- *"Craig-Geffen-Morse" approach*: This approach takes into account the impact of differences in vertical permeability and the ability of fluids to move.

Simulation Methods

- Numerical reservoir simulation models combine vertical heterogeneity and gravity effects to accurately simulate fluid flow and forecast the efficiency of vertical sweep.

8.2.2.4 Enhanced Oil Recovery (EOR) Techniques and Sweep Efficiency

Various EOR strategies have different effects on the efficiency of sweeping oil over the reservoir in both the horizontal and vertical directions. Below is a concise summary of how various EOR techniques impact the effectiveness of fluid displacement:

- Water flooding is a technique that enhances the effectiveness of sweeping a region, particularly in reservoirs that have a reasonably uniform composition. However, in reservoirs with varying properties, water has a tendency to flow along areas with high permeability, resulting in suboptimal distribution and utilization of the injected fluid.
- Polymer flooding is a technique that enhances the thickness of the water being injected, resulting in a decrease in the mobility ratio (M). This, in turn, improves the efficiency of both the horizontal and vertical movement of the injected water. The increased viscosity of the polymer solution enhances its ability to efficiently displace oil from zones with poorer permeability.
- Surfactant flooding is a technique that decreases the IFT between oil and water, hence enhancing the effectiveness of displacement. Surfactants

improve the efficiency of both horizontal and vertical displacement by lowering capillary forces.

- *CO₂ injection*: The process of injecting CO_2, particularly when it is able to mix evenly with oil, decreases the thickness of the oil and causes it to expand, hence enhancing the effectiveness of displacing the oil. Carbon dioxide (CO_2) also plays a crucial role in maintaining reservoir pressure, hence enhancing the efficiency of fluid displacement.
- Steam injection is a method that decreases the thickness of oil by using heat, which improves the efficiency of oil recovery by increasing the upward movement of oil. Steam-assisted gravity drainage (SAGD) optimally employs gravity to enhance the effectiveness of oil displacement in reservoirs with heavy oil.

Example: Calculation

Consider a reservoir with the following parameters:

- The mobility ratio (M) is equal to 0.5.
- The areal sweep efficiency (E_A) is estimated to be 0.8 based on simulation models.
- The vertical sweep efficiency (E_I) is estimated to be 0.7 based on analytical models.
- Calculate volumetric sweep efficiency (E_V)?

Solution:
From Equation (8.3), we have

$$\text{Overall volumetric sweep efficiency } (E_V) = E_A E_I$$

After substituting the values, $E_V = 0.56$.

Hence, the volumetric sweep efficiency stands at 56%, denoting that 56% of the pore volume in the reservoir has been effectively displaced by the injected fluid.

8.2.3   CLASSIFICATION OF **EOR** TECHNIQUES

The three primary classifications of EOR are thermal recovery, gas injection, and chemical injection. Figure 8.2 shows how the EOR method is categorized by the type of fluid that is injected, how the process is carried out, and the means of production.

8.2.3.1   Thermal Recovery

Thermal recovery techniques entail the application of heat to the reservoir in order to decrease the oil's viscosity, hence facilitating its movement toward the production wells.

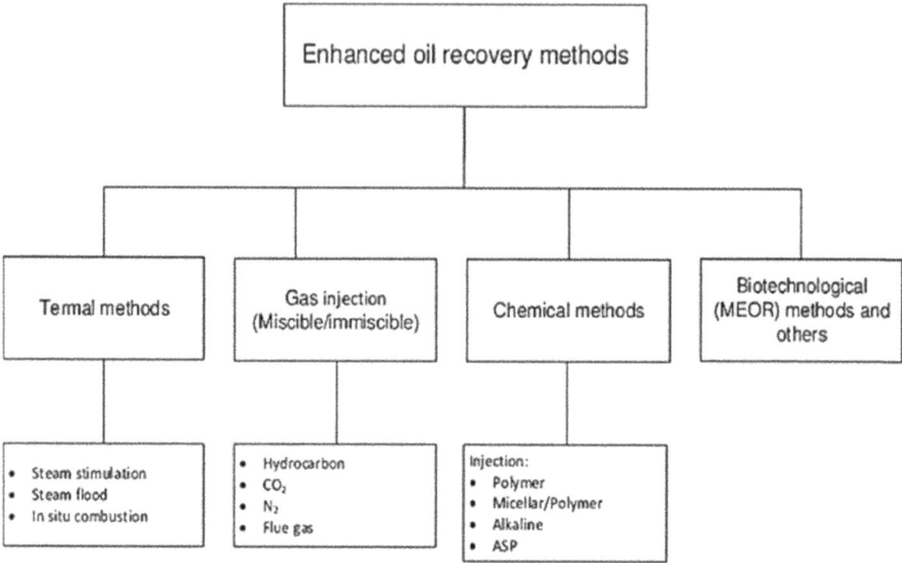

FIGURE 8.2 Available tertiary (enhanced oil recovery) techniques.[1]

a. Steam Injection

Steam injection is the predominant thermal EOR technique. The process entails injecting steam into the reservoir, which raises the temperature of the oil and reduces its viscosity. There are two primary categories:

Steam flooding: Steam flooding is the process of injecting steam continuously in order to displace the oil and push it toward production wells.

Cyclic steam stimulation (CSS) is a thermal recovery method used in the oil and gas industry. Referred to as "huff and puff," this technique entails the injection of steam into a well, followed by a soaking period, and subsequently extracting the oil.

b. In situ Combustion

It refers to a process when the combustion of a substance occurs in its original location or natural environment.

In situ combustion, also known as fire flooding, is a method that involves deliberately setting fire to a portion of the oil in the reservoir in order to produce heat. To maintain combustion, air or oxygen is introduced, which heats the surrounding oil and decreases its viscosity.

8.2.3.2 Gas Injection

Gas injection techniques entail the introduction of gases into the reservoir with the purpose of either sustaining pressure or blending with the oil to decrease its viscosity.

a. Miscible Gas Injection

Miscible gas injection involves the introduction of gases, such as carbon dioxide (CO_2), natural gas, or nitrogen, into the reservoir. The gases combine with the oil to create a unified state, decreasing the oil's thickness and causing it to expand, so facilitating its movement toward the production wells.

b. **Immiscible Gas Injection**

Immiscible gas injection involves the use of gases, such as nitrogen or flue gases, that do not mix with the oil. These gases are employed to sustain reservoir pressure and push the oil into the production wells.

8.2.3.3 Chemical Injection

Chemical injection methods entail the introduction of chemicals into the reservoir with the aim of enhancing oil recovery.

a. **Polymer Flooding**

Polymer flooding is a technique that entails injecting water-soluble polymers into the reservoir in order to enhance the thickness of the displacing water. This enhances the sweep efficiency and propels the oil toward the production wells.

b. **Surfactant Flooding**

Surfactant flooding is a process where surfactants are injected into the reservoir to decrease the IFT between oil and water. This enables the oil to move more smoothly toward the production wells.

c. **Alkaline Flooding**

Alkaline flooding is a process where alkaline chemicals, like sodium hydroxide, are injected into the reservoir. The chemicals undergo a reaction with the acidic components of the oil, resulting in the formation of surfactants in the same location. This process aids in lowering the tension between different surfaces and enhances the extraction of oil.

8.2.3.4 Advancing Enhanced Oil Recovery Techniques

Numerous nascent EOR methods are currently being investigated and in the process of being developed to further augment the extraction of oil. The following items are included:

a. **Microbial EOR (MEOR)**

Microbial EOR is the process of utilizing microorganisms to improve the extraction of oil. These microorganisms have the ability to generate gases, acids, or surfactants that aid in the movement of the trapped oil.

b. **Nanotechnology**

Nanotechnology utilizes nanoparticles (NPs) to enhance the process of oil recovery. These NPs can be engineered to decrease the thickness of oil, modify the ability of rocks to hold oil, or enhance the effectiveness of other EOR techniques.

8.2.3.4.1 *Advantages and Challenges of EOR Techniques*

- EOR techniques have the capacity to substantially augment the quantity of oil extracted from a reservoir, hence prolonging the lifespan of oil fields.

- EOR can yield significant economic advantages for both oil firms and the host governments by augmenting oil production.
- EOR offers environmental advantages by optimizing oil extraction from current fields, hence minimizing the necessity for additional drilling and exploration, and ultimately reducing the environmental effect of oil production.

8.2.3.4.2 Challenges

- *Expensive implementation*: Employing EOR approaches necessitates substantial investment in technology and infrastructure, resulting in high expenses.
- *Technical complexity*: EOR procedures are intricate and necessitate a profound comprehension of reservoir characteristics and fluid dynamics.
- *Environmental considerations*: Certain EOR techniques, including thermal recovery, might give rise to environmental consequences, such as the emission of greenhouse gases and the consumption of water.

8.3 LIMITATIONS OF AVAILABLE EOR TECHNIQUES

Conventional EOR techniques, such as thermal recovery, gas injection, and chemical flooding, have been widely employed to enhance the extraction of oil from reservoirs. Nevertheless, ultrasound-assisted nanofluid flooding holds the potential to overcome the constraints associated with these approaches. The limitations of conventional EOR techniques include the following thermal recovery methods.

8.3.1 STEAM INJECTION

Steam generation is energy-intensive, which means it requires a large amount of energy. This leads to higher operational expenses and a larger carbon footprint.

- *Reservoir damage*: High temperatures can damage the reservoir rock and its properties.
- *Restricted depth*: This method is only efficient in reservoirs that are not excessively deep, as the amount of heat loss increases as the depth grows.
- *Water usage*: Demands substantial amounts of water, which can be limited or costly.

Example: Heavy Oil Reservoirs

Steam injection is a widely employed technique in heavy oil reservoirs to decrease the viscosity of oil. Nevertheless, it is exceedingly energy-intensive and has the potential to cause harm to reservoirs and result in substantial water usage. On the other hand, ultrasound-assisted nanofluid flooding can improve the movement of oil without requiring high temperatures or significant water use.

8.3.2 Gas Injection Methods

8.3.2.1 Carbon Dioxide (CO_2) Injection

Operating conditions while handling and transporting CO_2 can pose significant challenges and incur high costs.

- Sweep efficiency is frequently compromised by gravity override and viscous fingering, resulting in suboptimal performance.
- *Corrosion*: Carbon dioxide (CO_2) has the potential to induce corrosion in pipes and equipment.

Example: Light Oil Reservoirs

CO_2 injection is employed in light oil reservoirs to enhance oil displacement. However, it frequently leads to suboptimal sweep efficiency and corrosion problems. Ultrasound-assisted nanofluid flooding offers a method of displacement that is more consistent and prevents the corrosion issues often linked to CO_2.

8.3.3 Chemical Flooding Methods

8.3.3.1 Polymer Flooding

- *Controlling viscosity*: It can be challenging to maintain the proper viscosity of polymer solutions over time and under reservoir conditions.
- Shear degradation refers to the process in which polymers undergo degradation when exposed to strong shear conditions in the reservoir, resulting in a decrease in their effectiveness.
- *Expense*: Polymers can have a high cost, and their effectiveness can be affected by reservoir conditions such as salinity and temperature.

8.3.3.2 Surfactant Flooding

- *Chemical stability*: Surfactants may undergo degradation within the reservoir, particularly when exposed to elevated temperatures or specific minerals.
- *Adsorption losses*: Surfactants have the ability to adhere to reservoir rocks, hence decreasing their accessibility for oil displacement.
- *Environmental impact*: Possible environmental issues associated with the utilization of substantial amounts of chemicals.

Example: Sandstone Reservoirs

Polymer and surfactant flooding techniques are employed in sandstone reservoirs to augment oil extraction. However, these approaches are plagued by problems such as chemical deterioration, losses due to adsorption, and expensive expenses. Utilizing stable NPs in ultrasound-assisted nanofluid flooding can lead to comparable or superior oil recovery outcomes, while avoiding the associated disadvantages.

8.4   ULTRASOUND-ASSISTED NANOFLUID FLOODING

Conventional EOR techniques, such as thermal recovery, gas injection, and chemical flooding, have many drawbacks, including energy usage, implications for the environment, chemical durability, and effectiveness in covering the entire oil reservoir. EOR methods are being extensively studied and utilized to optimize the retrieval of oil from current reservoirs. Out of these options, the method of combining nanotechnology with ultrasonic waves, referred to as *ultrasound-assisted nanofluid flooding*, offers a creative and novel approach. This method exploits the distinctive characteristics of NPs and the mechanical impacts of ultrasonic waves to enhance the effectiveness of oil displacement.

8.4.1   Fundamentals of Ultrasound-Assisted Nanofluid Flooding Nanofluids in EOR

Nanofluids are colloidal mixtures consisting of NPs, which have sizes ranging from 1 to 100 nm, dispersed in base fluids like water or oil. NPs can be synthesized using a diverse range of materials, such as metals (e.g., gold, silver), metal oxides (e.g., silica, alumina, titanium dioxide), and carbon-based materials (e.g., carbon nanotubes, graphene). Nanofluids possess distinct characteristics, including elevated surface area and reactivity, which render them well-suited for altering the parameters of reservoir fluids and rocks in order to augment oil recovery.

The extremely compact dimensions of nanoparticles and the resulting advantages they provide in the context of enhanced oil recovery (EOR) processes.

8.4.1.1   Ultra-Small Size of NPs

- The size of commonly used NPs such as SiO_2 (silica), TiO_2 (titanium dioxide), and Al_2O_2 (alumina) often falls within the range of 1 to 100 nm.
- The pore throat diameters in reservoir rocks often vary between a few nanometers and many micrometers.
- For example, sandstone reservoirs typically have pore sizes ranging from 1 to 10 µm, whereas carbonate reservoirs might have much greater pore sizes.

8.4.1.2   Advantage in Permeability

- *Minimized pore clogging*: Conventional chemical EOR agents have the potential to clog pores due to the presence of bigger molecules or aggregates, which can become trapped in the porous medium and subsequently decrease the permeability of the formation. This obstruction not only impedes the process of extracting oil but also leads to an increase in the expenses associated with injecting these fluids, as higher pressures are needed.
- NPs are able to move across pore spaces without creating a significant drop in permeability due to their extremely small size, resulting in efficient flow. The streamlined movement of fluids within the reservoir improves the overall efficiency of the EOR process.

Example: Silica Nanoparticles

Silica NPs, because to their diminutive size, can traverse sandstone reservoirs without obstructing the pore throats. This mitigates the decrease in formation permeability, a frequent problem encountered with bigger chemical molecules or polymer flooding agents.

8.4.1.3   Enhanced Sweep Efficiency

- *Micro and nano pores*: Traditional injection fluids may not effectively penetrate smaller pores due to their larger molecular size or because of high IFT between the injection fluid and oil. This constraint diminishes the effectiveness of the sweeping process and results in a substantial volume of oil being left behind.
- NP*s' penetration*: Owing to their extremely small size, NPs are capable of infiltrating micro and nano pores that conventional fluids are unable to reach. This capability enables NPs to come into contact with a greater number of oil-rich areas within the reservoir, hence enhancing the overall efficiency of oil recovery.
- Macroscopic sweep efficiency refers to the effectiveness of displacing fluids in a reservoir at a large scale. The macroscopic sweep efficiency quantifies the effectiveness of the injected fluid in displacing oil across the whole reservoir. NPs enhance efficiency by effectively accessing and mobilizing oil inside even the tiniest pores.
- NPs have the ability to decrease the IFT between oil and water, hence improving the efficiency of displacement. The decrease in IFT facilitates the movement of the trapped oil droplets and enhances the overall recovery process.

Example: Titanium Dioxide Nanoparticles

Titanium dioxide NPs (TiO_2 NPs) have been shown to modify the wettability and decrease the IFT in carbonate reservoirs. This enables them to penetrate smaller pores more effectively and enhance the overall sweep efficiency of the reservoir.

8.4.1.4   Reduced Chemical Usage and Environmental Impact

- *Environmental and economic benefits*: Reduced chemical usage: Conventional EOR techniques typically necessitate substantial amounts of chemicals, resulting in high expenses and potential environmental hazards. Conversely, NPs can make substantial enhancements in EOR with significantly less amounts, hence cutting expenses and minimizing environmental consequences.
- The utilization of environmentally friendly NPs, such as silica, in the EOR process guarantees both efficiency and sustainability, thereby reducing the adverse environmental effects linked to chemical flooding.

Example: Alumina Nanoparticles

Alumina NPs (Al_2O_3 NPs) have been employed in several reservoirs to augment oil recovery. Due to their tiny size and capacity to change wettability and decrease

IFT, these substances use less chemical volume than traditional approaches, resulting in a more environmentally friendly process.

8.4.2 Mechanisms for Nanofluid-Assisted EOR

A nanofluid is a stable suspension consisting of a base fluid and NPs that have an average size smaller than 100 nm. The base fluid may consist of any liquid, including oil, water, or gas. Nanofluids, which consist of water or brine solution with added NPs, are commonly employed to enhance water flooding recovery. The EOR processes of nanofluids have been extensively studied and are a novel approach in the current scenario. These mechanisms mostly involve:

- disjoining pressure,
- pore channels plugging,
- viscosity increase of injection fluids,
- IFT reduction,
- wettability alteration,
- prevention of asphaltene precipitation.

Figure 8.3 displays the schematic of the EOR mechanisms of nanofluids.

8.4.2.1 Disjoining Pressure

Disjoining pressure (Π) refers to the pressure disparity across a thin liquid film resulting from the influence of intermolecular and surface forces. It has a crucial impact on the durability and thickness of thin liquid layers in porous materials, like those present in oil reservoirs. The pressure can be affected by van der Waals forces, electrostatic forces, and structural forces. Upon contact with the oil phase, the NPs

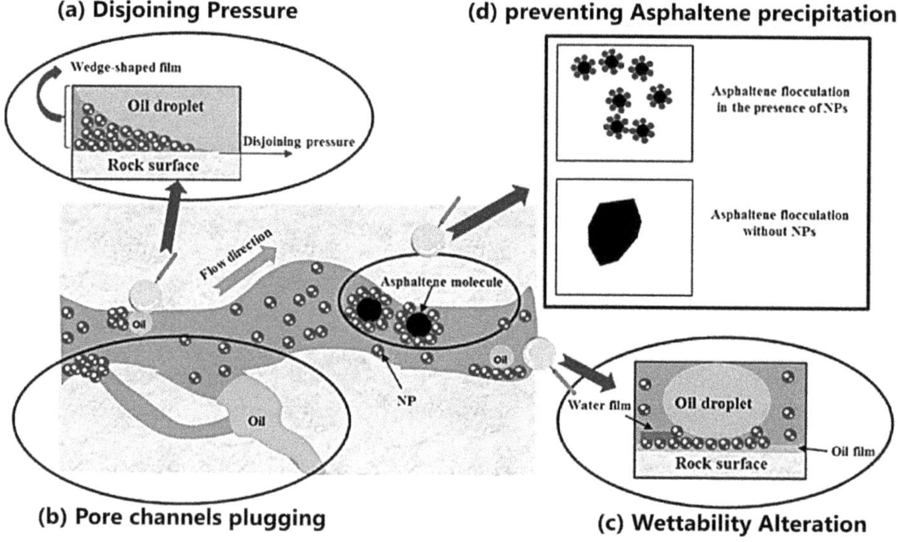

FIGURE 8.3   Mechanism for nanofluid-assisted EOR technique.[2]

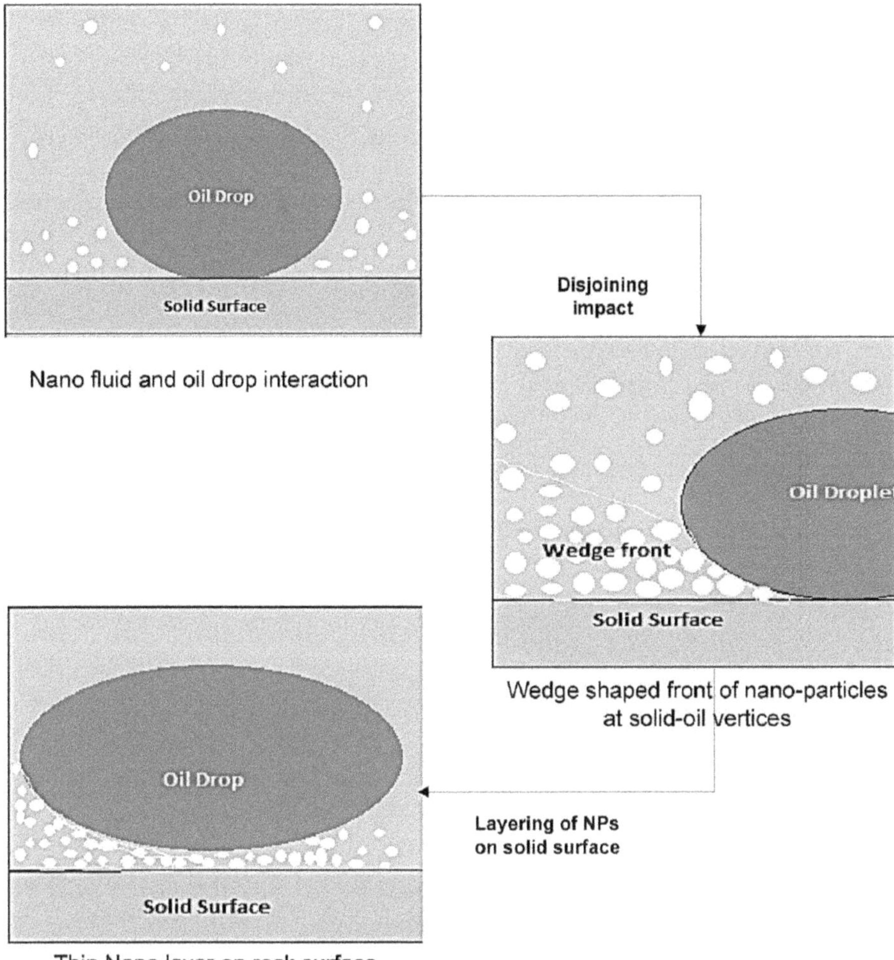

FIGURE 8.4 Self-assembled nanofluid liquid film with a wedge-shaped structure.[3]

in the nanofluids can arrange themselves into a self-assembled film with a wedge-shaped structure, as depicted in Figure 8.4. The wedge film functions by effectively isolating the oil droplets from the rock surface, resulting in a higher recovery of oil compared to typical injection fluids. The presence of a pressure known as structural disjoining pressure leads to the formation of a wedge-shaped film.

8.4.2.1.1 Key Concepts
Formation of Thin Liquid Film:

- A thin liquid film emerges at the boundary between two interfaces, such as oil and water or oil and rock surfaces.
- These coatings are present in oil reservoirs, existing between oil droplets and the adjacent water or rock surfaces.

Intermolecular Forces:

- van der Waals forces refer to the intermolecular forces that can be either attractive or repulsive in nature.
- Electrostatic forces refer to the attractive or repulsive forces that exist between charged particles.
- Structural forces refer to the forces that arise from the specific organization and composition of molecules.

Mechanism of Disjoining Pressure in Nanofluids:

- *NP introduction*: NPs are injected into the reservoir via the injection fluid.
- The NPs can consist of metal oxides, silica, or other materials specifically engineered to interact with the surfaces of fluids and rocks.

Surface Adsorption:

- NPs adhere to the surfaces of the oil droplets and rock surfaces. This process of adsorption alters the way the liquid film and the surfaces interact with one other.

Increase in Disjoining Pressure:

- NPs elevate the disjoining pressure by augmenting the repulsive forces within the thin liquid film.
- Elevated disjoining pressure (Π) enhances the stability of the thin liquid layer, inhibiting its rupture and the merging of oil droplets.

Thin Film Stabilization:

- The stabilized thin layers hinder the merging of oil droplets, hence preserving their separate condition.
- Stabilization is essential in order to release trapped oil by creating uninterrupted paths for oil to move toward the production well.

Improved Oil Mobilization:

- Stabilized thin coatings facilitate the enhanced mobility of oil droplets within the porous rock matrix.
- This optimized mobilization enhances the overall effectiveness of the EOR process, leading to a higher amount of oil being recovered.

8.4.2.2 Pore Channel Plugging

The obstruction of pore channels can have a substantial effect on the effectiveness of EOR methods. Plugging can occur through two main mechanisms as shown in Figure 8.5: mechanical trapping and log-jamming. Below, both methods are elucidated comprehensively:

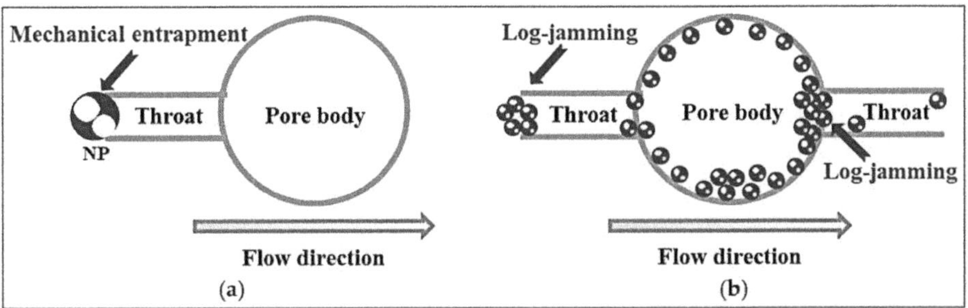

FIGURE 8.5 Pore channel plugging mechanism: (a) mechanical trapping and (b) log-jamming.[2]

8.4.2.2.1 Mechanical Entrapment (Trapping)

Mechanical entrapment occurs when NPs or other solid particles physically obstruct the pore channels. This phenomenon arises when the particles are incapable of traversing the pore throats, which are the most constricted sections of the pore channels.

8.4.2.2.1.1 Operational Process
Characteristics of NP Size and Pore Throat Diameter:

- The relative size of the NPs in relation to the diameter of the pore throat is of utmost importance.
- If the size of the NPs is equal to or greater than the diameter of the pore throat, they become stuck and obstruct the channel.

Flow Dynamics:

- As the injection fluid transports NPs through the permeable material, certain NPs get trapped in the narrow passages of the pores.
- The obstruction can be of a transient or enduring nature, contingent upon the fluid dynamics and the interplay between the particles and the walls of the pore.

Effects:

- Mechanical entrapment diminishes the permeability of the porous material in the obstructed areas.
- As a result, there is an elevated resistance to the flow, causing the injection fluid to be redirected into other zones that have lower permeability.

8.4.2.2.2 Log-Jamming:
- Log-jamming is the phenomenon where a significant quantity of NPs gather and cluster together in the pore channels, resulting in a blockage. This mechanism is referred to as logjam, which occurs when logs accumulate and impede the movement of water in a river.

8.4.2.2.2.1   Mechanism
Nanoparticle Aggregation:

- NPs have the ability to come together and form aggregates as a result of van der Waals forces, electrostatic interactions, or other attractive forces.
- Aggregation enhances the overall dimensions of the particles, hence facilitating their ability to induce obstructions.

Flow-Induced Accumulation:

- As the fluid passes through the porous material, clustered NPs have a tendency to gather at narrow points within the channels of the pores.
- This buildup results in a gradual accumulation, ultimately resulting in a logjam.

Effects:

- Log-jamming greatly decreases the permeability of the impacted areas, akin to being mechanically trapped.
- It results in a redistribution of the injection fluid flow, which improves the effectiveness of sweeping by directing the fluid into areas with lower permeability.

8.4.2.3   Viscosity Increases of Injection Fluids

EOR can greatly enhance the effectiveness of the oil displacement process by increasing the viscosity of the injection fluid. Nanofluids, consisting of NPs dispersed in a base fluid, can be developed to enhance the viscosity of injectable fluids.

8.4.2.3.1   Importance in EOR

The primary goal in EOR is to improve the displacement of oil within the reservoir. Enhancing the viscosity of the injection fluid improves the control of fluid movement, reducing the likelihood of the injected fluid bypassing zones with high oil content and enhancing the overall efficiency of the sweep.

8.4.2.3.2   Mechanisms of Viscosity Increase
Nanoparticle Interactions:

- The introduction of NPs into a base fluid leads to interactions between the NPs themselves and the molecules of the fluid. These interactions result in an augmentation of the fluid's viscosity.
- *Hydrodynamic interactions*: NPs have an impact on the flow and motion of fluid molecules, resulting in increased resistance to flow and higher viscosity.

Particle Size and Shape:

- *Size*: Decreasing the size of NPs results in a greater surface area available for interaction with the molecules of the base fluid, leading to a substantial rise in viscosity.
- NPs exhibit different shapes (spherical, rod-like, plate-like), which significantly impact their interactions with both the fluid and other NPs. Particles with a rod-like or plate-like shape exhibit more resistance to flow in comparison to particles with a spherical shape.

Nanoparticle Concentration:

- Higher concentration: Augmenting the concentration of NPs in the fluid yields a greater number of interactions and collisions between particles, resulting in an elevated viscosity.
- *Optimal concentration*: Beyond a certain concentration, the viscosity may increase to a point where it becomes counterproductive, potentially causing problems with injectivity and flow through the reservoir.

Surface Chemistry and Coatings:
Surface modifications involve the application of different compounds onto NPs in order to improve their stability and interaction with the base fluid.

Dispersants are used to preserve a consistent suspension of NPs, preventing them from clumping together and guaranteeing a uniform rise in viscosity.

8.4.2.3.3   Benefits of Increased Viscosity
Enhance Enhanced Mobility Control:

- *Mobility ratio*: The mobility ratio (M) represents the relationship between the mobility of the displacing fluid and the mobility of the displaced fluid. A lower mobility ratio signifies superior displacement efficiency.
- *Fingering control*: Increased viscosity of the injection fluid mitigates the occurrence of viscous fingering, a phenomenon in which the injected fluid penetrates the reservoir unevenly, bypassing areas rich in oil.

Improved Sweep Efficiency:

- Enhancing the viscosity of the injection fluid results in a more consistent displacement front, which in turn leads to a more comprehensive sweep of the reservoir.
- *Enhanced contact*: The increased thickness of the substance improves its ability to stay in contact with the oil, making it easier to move and recover.

Better Conformance Control:

- High-permeability zones refer to areas in heterogeneous reservoirs where fluids with higher viscosity tend to enter low-permeability zones. This helps to prevent early breakthrough in channels with high permeability.

- Flow redistribution occurs when high-permeability channels are blocked, causing the injection fluid to be rerouted toward low-permeability zones that contain a high concentration of oil. This process enhances the overall recovery of oil.

8.4.2.4 IFT Reduction

IFT refers to the tension exerted at the boundary between two fluids that are immiscible in nature, such as oil and water. It is a crucial factor in determining the distribution and flow of fluids within porous materials, such as oil reservoirs. Elevated IFT can impede the process of oil recovery by generating potent capillary forces that ensnare oil within the rock formation of the reservoir. Lowering IFT helps to improve the movement and displacement of oil that is trapped, which is a crucial mechanism in approaches used to enhance oil recovery (EOR).

IFT Reduction Mechanism Utilizing Nanoparticles (NPs):

- NPs have the ability to significantly decrease the IFT between oil and injection fluids, such as water or brine, through many mechanisms.
- Adsorption occurs when NPs move toward and adhere to the interface between oil and water, causing changes in the characteristics of the contact.
- Adsorption diminishes the cohesive forces within the oil phase and the adhesive forces between oil and water molecules, resulting in a decrease in the IFT.

Modification of Interfacial Rheology:

- NPs alter the rheological characteristics of the contact, including its elasticity and viscosity.
- This modification aids in the stabilization of emulsions and decreases the energy needed to change the interface, resulting in a reduced IFT.

Surface Modification of Nanoparticles

- NPs can undergo surface modification through the application of surface-active chemicals or surfactants. This process involves coating the NPs, which improves their capacity to migrate to and stabilize the interface between oil and water.
- These altered NPs can offer a more substantial decrease in IFT in comparison to NPs that have not been modified.

Electrostatic and van der Waals Interactions:

- NPs can engage in interactions with oil and water molecules via electrostatic and van der Waals forces.
- These interactions diminish the strength of the forces at the interface, leading to a decrease in IFT.

8.4.2.4.1   Benefits of IFT Reduction in EOR
Enhanced Oil Mobilization:

- Decreasing IFT diminishes the capillary forces that confine oil within the reservoir rock, facilitating the displacement of trapped oil by the injected fluid.
- As a result, there is enhanced mobilization and recuperation of oil.

Enhanced Sweep Efficiency:

- By decreasing the IFT, the fluid that is injected can efficiently traverse the reservoir, coming into contact with a greater volume of oil and so diminishing the quantity of remaining oil.
- As a result, this leads to a more effective displacement procedure and increased rates of oil recovery.

Emulsion Stabilization:

- Decreasing the IFT helps stabilize emulsions when oil is dispersed in water or water is dispersed in oil. This can facilitate the transportation of oil through the reservoir and production system.
- Stabilized emulsions have the ability to hinder the merging of oil droplets, therefore preserving their ability to move.

8.4.2.5   Wettability Alteration

Wettability refers to the inclination of one fluid to stick to the surface of a rock, while in competition with another fluid that cannot mix with it. Wettability, in the context of oil reservoirs, governs the manner in which water and oil engage with the surface of the rock. It is a crucial factor that has a significant impact on capillary pressure, fluid saturation, and relative permeability, all of which ultimately affect the efficiency of oil recovery.

Importance of Wettability in Oil Recovery Capillary Pressure:

- Wettability influences the capillary forces inside the rock formation of the reservoir. Water-wet circumstances result in elevated capillary pressure, facilitating the displacement of oil by water.
- Under oil-wet circumstances, the capillary pressure decreases, which hinders the displacement of oil by water.

Fluid Saturation:

- Wettability governs the spatial arrangement of oil and water within the rock's pores. Rocks that are wet with water have a greater capacity to hold water in their smaller pores, but rocks that are wet with oil have a greater capacity to hold oil.
- Modifying the wettability can manipulate the saturation levels, hence impacting the quantity of oil that can be extracted.

Relative Permeability:

- Wettability affects the level of ease at which oil and water pass through the rock. Water-wet conditions promote the movement of water, whereas oil-wet conditions promote the movement of oil.
- Modifying the surface properties to control wettability can enhance the efficiency of oil extraction by optimizing the relative permeability.

Ways in which wettability can be changed by utilizing nanofluids are as follows:

- Nanofluids, consisting of NPs suspended in a base fluid, have been acknowledged for their ability to modify the wettability of reservoir rocks. The main mechanisms consist of:

Nanoparticle Adsorption:

- *Surface coating*: NPs adhere to the rock surface, creating a layer that can modify the surface energy and wettability.
- The hydrophilic/hydrophobic balance of NPs can be adjusted to modify the surface characteristics of rocks, making them either more water-wet (hydrophilic) or oil-wet (hydrophobic).

Surface Energy Modification:

- *Chemical interaction*: NPs have the ability to chemically interact with the surface of the rock, causing changes to its surface energy.
- Functionalization refers to the process of modifying the surface of NPs by attaching specific chemical groups. This modification might improve the way the NPs interact with the surface of rocks, leading to a change in their wettability.

Capillary Force Adjustment:

- NPs have the potential to decrease the IFT between oil and water, which makes it easier for wettability to be altered.
- The addition of NPs can modify the contact angle between the rock surface and the fluids, indicating a change in wettability.

Surface Roughness:

- Nano structuring refers to the process of creating nanostructures on the surface of a rock using NPs. This alters the roughness of the rock surface, which in turn affects its wettability.
- Increased NP adsorption leads to an enhanced surface area, which in turn enhances the impacts of wettability change.

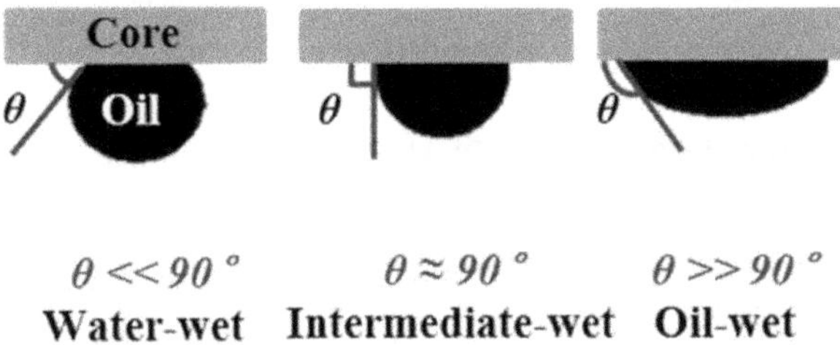

FIGURE 8.6 Schematic diagram of an oil-nanofluid system wettability alteration in terms of contact angle measurement.[2]

8.4.2.6 Wettability Alteration in Nanofluid-Assisted EOR in Terms of Contact Angle

8.4.2.6.1 Contact Angle

The contact angle is a quantitative indicator of how easily a liquid spreads or adheres to a solid surface. The contact angle is the angle created at the point of intersection between a liquid interface and a solid surface. The contact angle offers valuable information regarding the wetting characteristics of a surface, indicating whether it is water-wet, oil-wet, or intermediate-wet as shown in Figure 8.6.

- A *water-wet surface* is characterized by a narrow contact angle ($\theta < 90°$), indicating that water readily spreads across the surface. This suggests a high level of adhesion between the water and the rock surface.
- An *oil-wet surface* is characterized by a contact angle greater than 90°, indicating that water forms beads on the surface. This suggests that the adhesion between water and the rock surface is weak, while the adhesion between oil and the rock surface is stronger.
- *Intermediate-wet*: A contact angle that is around 90° suggests a state of wetting that is well-balanced.

Beneficial Effects of Altering Wettability in Enhanced Oil Mobilization for EOR Application:

- The reduction of the contact angle, which is accomplished through the adsorption of hydrophilic NPs, makes it easier for oil droplets to detach themselves from the surface of the rock, which in turn increases the amount of oil that is mobilized.[4]
- Lower contact angles result in an increase in capillary forces, which contribute to the displacement of oil from the pores of the rock.
- *Uniform displacement*: A rock surface that is more water-wet results in a more uniform displacement front, which in turn reduces the amount of oil-rich zones that escape the front.
- Lower contact angles reduce the amount of viscous fingering that occurs, which in turn improves the overall sweep efficiency to a greater extent.
- Adjusting the wettability to circumstances that are more water-wet optimizes the relative permeability of oil and water, which in turn increases the flow of oil toward production wells. This results in a balanced flow.

Case Study 1

EXAMPLE CALCULATION AND EXPLANATION

Let us examine a rock that is oil-wet, meaning it has a high affinity for oil, and has an initial contact angle of 120°. The contact angle is determined to be 45° (water-wet state) after the injection of a nanofluid containing hydrophilic NPs.

Contact angle before nanofluid injection: $\theta_{initial} = 120°$

Contact angle after nanofluid injection: $\theta_{final} = 45°$

$$\Delta\theta = \theta_{initial} - \theta_{final}$$

Change in contact angle:

$$\Delta\theta = 75°$$

The substantial decrease in the contact angle signifies a successful modification of the rock's wettability from being inclined toward oil to being inclined toward water, hence improving the efficiency of oil displacement and recovery.

8.5 LIMITATIONS OF NANOFLUID-INDUCED ENHANCED OIL RECOVERY (EOR)

Although nanofluid-induced EOR presents various benefits, its application is hindered by significant constraints and challenges. It is crucial to carefully evaluate these limits in order to maximize the efficiency and financial feasibility of nanofluid EOR approaches.[5]

Cost of Nanoparticles

- *Expensive production costs*: The process of synthesizing and functionalizing NPs might incur high expenses. The process of manufacturing a significant quantity of NPs with consistent size and good quality is expensive.
- *Economic viability*: The ratio of costs to benefits must be advantageous. If the EOR does not provide sufficient economic benefits to offset the extra expenses, the use of nanofluid EOR may not be financially feasible.[6]

Nanoparticle Stability and Aggregation

- *Agglomeration*: NPs have a tendency to come together in the fluid, particularly in reservoirs where there are high levels of salinity and temperature. This aggregation has the potential to diminish their potency and obstruct pore spaces.
- Ensuring the stability of NP suspension in the injection fluid for long durations is a difficult task. Chemical additives, specifically dispersants, may be necessary to enhance the stability of the NPs, hence increasing both the complexity and cost of the process.[7]

Concerns Regarding the Environment and Human Health

- NPs can have adverse effects on both the environment and human health, potentially causing toxicity. It is imperative to conduct a comprehensive assessment of the impact of NPs on both the ecosystem and human health.

- Ensuring regulatory compliance for the use of nanofluids involves adhering to environmental rules and acquiring the required licenses, which may be a time-consuming and expensive process.[8]

Reservoir Heterogeneity

- *Challenging reservoir conditions*: Reservoirs frequently exhibit heterogeneity, characterized by diverse geological characteristics, fluid compositions, and temperatures. The intricate nature of nanofluids can provide challenges in accurately forecasting their behavior and how they interact with the reservoir rock and fluids.
- NPs exhibit selective penetration by favoring entry into regions with high permeability, while bypassing regions with low permeability where oil is held. This can impede the overall efficacy of the EOR process.

Injectivity Issues

- *Formation damage*: The introduction of NPs might result in formation damage by obstructing the narrow passages within the reservoir rock, hence decreasing its ability to allow fluid flow and injection.
- *Pressure buildup*: The elevated viscosity of the nanofluid can lead to greater injection pressures, which can jeopardize the reservoir's integrity and the injection equipment.

Interaction with Reservoir Fluids and Rock Unforeseeable Interactions

- The interplay of NPs, reservoir fluids (oil, water, gas), and rock can be intricate and unforeseeable. These interactions have the potential to influence the stability and efficacy of the nanofluid.
- NPs can adhere to the surface of the rock or be trapped in the small crevices within it, which decreases the amount of active particles that can be used for EOR operations.

Field-Scale Implementation Challenges

- *Scale-up challenges*: Findings obtained in the laboratory may not always be directly applicable to large-scale operations in the field. Implementing the widespread usage of nanofluids in large reservoirs poses logistical and operational difficulties.
- Monitoring and control are essential for ensuring the optimal operation of the nanofluid injection process. This necessitates cutting-edge technology and specialized knowledge.

Uncertainty in Long-Term Effects

- The long-term performance and stability of nanofluids in the reservoir are little comprehended. Additional research is required to explore the potential long-term impacts on reservoir features and the extraction of oil.

- The worry lies in the destiny of residual NPs that remain in the reservoir following EOR procedures. If these particles were to migrate to other locations, they have the potential to modify reservoir features or create environmental problems.

8.6 SUMMARY

Nanofluid-induced EOR offers a promising approach to improve oil extraction by utilizing the distinct characteristics of NPs.

- The presence of NPs in nanofluid-assisted EOR leads to an elevation in disjoining pressure. This elevation helps to sustain thin liquid films in the reservoir, limiting the merging of oil droplets and improving their movement.
- Mechanical entrapment arises from the obstruction of pore throats by individual NPs, while log-jamming arises from the accumulation and aggregation of NPs, leading to the formation of larger blockages.
- Enhancing the viscosity of injection fluids through the use of nanofluids is a crucial approach in EOR.
- Reducing the IFT is beneficial for stabilizing emulsions where oil is mixed in water or water is scattered in oil. This can enhance the transportation of oil through the reservoir and production system.
- The contact angle is decreased by the adsorption of hydrophilic NPs, facilitating the detachment of oil droplets from the rock surface and consequently enhancing the mobilization of oil.

REFERENCES

1. Karović-Maričić, V., Leković, B., & Danilović, D. (2014). Factors influencing successful implementation of enhanced oil recovery projects. *Podzemni Radovi, 25*, 41–50.
2. Sun, X., Yanyu, Z., Guangpeng, C., & Zhiyong, G. (2017). Application of nanoparticles in enhanced oil recovery: a critical review of recent progress. *Energies, 10*(3), 345. https://doi.org/10.3390/en10030345
3. Chandio, T. A., Muhammad, A. M., Khalil Rehman, M., Ghulam, A., & Ghazanfer Raza, A. (2021). Enhanced oil recovery by hydrophilic silica nanofluid: experimental evaluation of the impact of parameters and mechanisms on recovery potential. *Energies, 14*(18), 5767. https://doi.org/10.3390/en14185767
4. Kumar, N., & Sonawane, S. S. (2018). Convective heat transfer of metal oxide-based nanofluids in a shell and tube heat exchanger. In *Conference Proceedings of the Second International Conference on Recent Advances in Bioenergy Research: ICRABR* 2016 (pp. 183–192). Singapore: Springer.
5. Kumar, N., & Sonawane, S. S. (2016). Experimental study of Fe_2O_3/water and Fe_2O_3/ethylene glycol nanofluid heat transfer enhancement in a shell and tube heat exchanger. *International Communications in Heat and Mass Transfer, 78*, 277–284.
6. Nishant, K., & Sonawane, S. S. (2016). Influence of CuO and TiO_2 nanoparticles in enhancing the overall heat transfer coefficient and thermal conductivity of water and ethylene glycol based nanofluids. *Research Journal of Chemistry and Environment, 20*, 8.

7. Khedkar, R. S., Shrivastava, N., Sonawane, S. S., & Wasewar, K. L. (2016). Experimental investigations and theoretical determination of thermal conductivity and viscosity of TiO_2–ethylene glycol nanofluid. *International Communications in Heat and Mass Transfer, 73*, 54–61.
8. Sonawane, S. S., Khedkar, R. S., & Wasewar, K. L. (2015). Effect of sonication time on enhancement of effective thermal conductivity of nano TiO_2–water, ethylene glycol, and paraffin oil nanofluids and models comparisons. *Journal of Experimental Nanoscience, 10*(4), 310–322.

9 Application of Cavitation-Induced Nanofluids in Wastewater Treatment

9.1 INTRODUCTION TO WASTEWATER TREATMENT

In recent years, the pollution of water has caused substantial environmental problems, mostly due to the presence of both organic and inorganic pollutants. One significant cause for concern, in addition to the existence of harmful pollutants, is the large volume of waste products produced. This is due to the extensive use of water in various industrial, agricultural, and domestic activities. The worries are exacerbated by the presence of numerous novel contaminants, such as medicines or pesticides, which have been identified in aquatic systems as a result of the shortcomings of traditional technology. Pharmaceuticals utilized for various purposes by humans and in animal husbandry are expelled from the body in feces and/or urine, either in their original form or as byproducts of metabolism. Consequently, they can readily infiltrate the aquatic system through both traditionally treated and untreated effluent discharge. Pesticide compounds and/or their intermediates are commonly found in industrial effluent and processed water from agricultural fields due to the extensive use of pesticides in agricultural activities, with not all of them being fully utilized.

The standard methods of wastewater treatment are unable to fully convert the emergent pollutants into carbon dioxide and water. Consequently, efforts are focused on developing and implementing advanced wastewater treatment technologies and processes in order to comply with stringent environmental laws. Advanced oxidation processes (AOPs) are effective methods for eliminating harmful pollutants, thanks to their high-efficiency and uncomplicated reactor architecture. AOP utilizes the production of extremely reactive hydroxyl radicals (OH) with the purpose of breaking down contaminants. The present study focuses on many facets of HC, which is a type of AOP used in wastewater treatment.

9.2 APPLICATION OF CAVITATION-INDUCED NANOFLUIDS IN WASTEWATER TREATMENT

Cavitation-induced nanofluids are a state-of-the-art solution in the field of wastewater treatment, providing a groundbreaking method for tackling the worldwide issue of water pollution. The rising levels of industrialization and urbanization have led to an increase in water pollution, which requires the use of advanced treatment

DOI: 10.1201/9781003594949-9

technologies to guarantee the supply of clean and safe water. Traditional techniques of treating wastewater, while they work well, typically have drawbacks in terms of how well they work, how much they cost, and their influence on the environment. Integrating cavitation phenomena with nanotechnology in this context offers a promising opportunity to improve the effectiveness of wastewater treatment procedures.

9.2.1 Understanding Cavitation

Cavitation is the occurrence of small vapor-filled cavities or bubbles due to rapid pressure changes in a liquid. These bubbles undergo rapid expansion and contraction, resulting in the release of highly concentrated energy in the form of shock waves and micro-jets. Cavitation, through the release of energy, can cause physical, chemical, and mechanical changes in the surrounding medium. This makes cavitation a valuable tool in several industrial applications, such as wastewater treatment. In this subject, the main categories of cavitation that are significant are HC and acoustic cavitation. Each of these types has unique mechanisms and applications.

- HC refers to the phenomenon that arises when a liquid passes through a narrow passage or across a surface, resulting in a decrease in pressure and the creation of vapor bubbles. When the bubbles collapse, they produce strong shear forces and concentrated heating, which can degrade contaminants and improve mixing.
- Acoustic cavitation is the result of the generation and collapse of bubbles caused by high-frequency ultrasonic waves. Acoustic cavitation involves the formation of cavities, through the use of ultrasonic waves at frequencies ranging from 16 kHz to 2 MHz. This method induces localized elevated temperatures and pressures, facilitating the decomposition of organic pollutants and the destruction of microbial cells.

HC reactors (Figure 9.1) are being considered as a more efficient alternative to acoustic cavitation due to some drawbacks of acoustic cavitation, including lower energy efficiency, limited possibilities for scaling up, and high operational costs. In the context of wastewater treatment, numerous studies have noted that HC alone does not achieve a satisfactory level of mineralization. This is because the production of oxidizing radicals is limited, and combining HC with other AOPs helps to enhance the oxidation capacity.

9.2.2 Hybrid Methods Based on Hydrodynamic Cavitation for Effluent Treatment

9.2.2.1 The Combination of Hydrodynamic Cavitation and Hydrogen Peroxide

The primary mechanism responsible for the breakdown of a pollutant by HC is the generation and subsequent reaction of •OH. H_2O_2 serves as an oxidizing agent to produce hydroxyl radicals, which can enhance the degradation process by HC treatment.

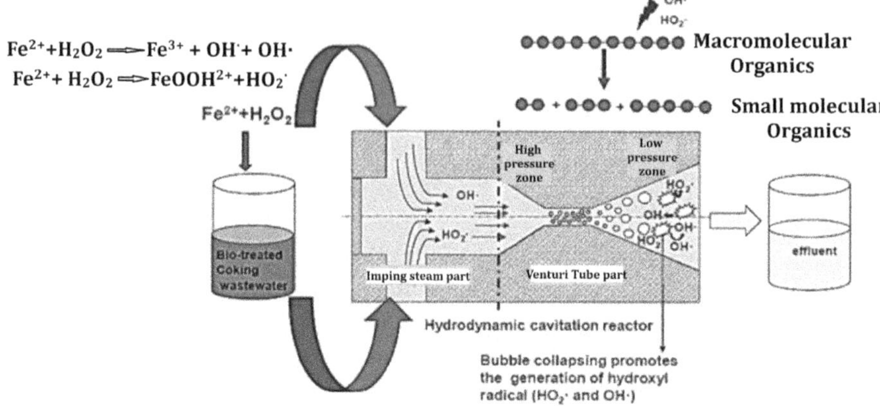

FIGURE 9.1 Schematic representation of hydrodynamic cavitation reactor for effluent treatment.[1]

It should be emphasized that H_2O_2 can also function as a scavenger, particularly at greater concentrations, resulting in a decrease in the level of degradation. The cavitation effect in the HC reactor leads to the dissociation of H_2O_2 into OH radicals. This occurs as a result of the conditions developed, including local hot spots with high temperature and pressure. The subsequent sequence of reactions takes place during the simultaneous application of HC and H_2O_2.

$$H_2O_2 \rightarrow OH + O\dot{H} \tag{9.1}$$

$$O\dot{H} + H_2O_2 \rightarrow HO_2 + H_2O \tag{9.2}$$

$$O\dot{H} + HO_2 \rightarrow O_2 + H_2O \tag{9.3}$$

$$HO_2 + H_2O_2 \rightarrow OH + H_2O + O_2 \tag{9.4}$$

$$\text{Effluent} + OH \rightarrow CO_2 + H_2O + \text{Degradation intermediates} \tag{9.5}$$

9.2.2.2 The Combination of Hydrodynamic Cavitation and Ozone

Ozonation is an efficient treatment method, because it has a high oxidation potential of 2.08 V, which allows it to effectively break down contaminants such as dyes and pesticides. The synergy of HC with ozone can be highly efficient. This is because the cavitation effect facilitates the dissociation of ozone into molecular O_2 and nascent oxygen (O), which then combine with water to create •OH. The subsequent sequence of reactions takes place with the simultaneous application of HC and ozone (O_3).

$$H_2O \rightarrow \dot{H} + \dot{OH} \tag{9.6}$$

$$O_3 \rightarrow O_2 + O \tag{9.7}$$

$$O + H_2O \rightarrow 2OH \tag{9.8}$$

$$\text{Effluent} + OH \rightarrow CO_2 + H_2O + \text{Degradation intermediates} \tag{9.9}$$

The Combined Effects of HC and the Ozone Process

- The utilization of HC with ozone treatment synergistically exploits the advantages of both processes, leading to heightened degradation of pollutants and enhanced efficiency in treatment. The primary synergistic effects encompass:
- Cavitation induces turbulence and micro-mixing, hence enhancing the mass transfer of ozone by increasing the interfacial contact area between ozone and the liquid. This enhances the solubility and dispersion of ozone, hence increasing its availibility for oxidation processes.
- The process of cavitation can create intense circumstances that promote the breakdown of ozone into highly reactive oxygen species (ROS), including •OH. This enhances the oxidative capacity of the therapy process.
- Cavitation exerts mechanical forces that can dismantle intricate chemical molecules and destabilize microbial cell walls, rendering contaminants more vulnerable to oxidative destruction by ozone.
- *Enhanced destruction kinetics*: The synergistic effect of cavitation and ozone expedites the oxidation process of pollutants, resulting in more rapid and thorough destruction. This decreases the amount of time needed for effective treatment.

9.2.2.3 The Combination of Hydrodynamic Cavitation and Fenton

The integration of HC and the Fenton process offers a promising solution for improving the efficiency and effectiveness of wastewater treatment. This novel approach capitalizes on the combined effects of cavitation and AOPs to more efficiently degrade pollutants, providing a feasible route toward achieving cleaner and safer water.

9.2.2.3.1 Fenton Process

The Fenton process is an AOP that uses H_2O_2 and ferrous iron (Fe^{2+}) to produce •OH. These radicals are extremely reactive and may break down various types of organic contaminants. The procedure encompasses the subsequent pivotal reactions:

$$Fe^{2+} + H_2O_2 \rightarrow Fe^{3+} + {}^{\bullet}OH + OH^- \tag{9.10}$$

$$Fe^{3+} + H_2O_2 \rightarrow Fe^{2+} + HO_2^{\bullet} + H^+ \tag{9.11}$$

The $^{\bullet}OH$ produced during the Fenton process exhibit exceptional efficacy in oxidizing and decomposing organic pollutants into less complex and less detrimental substances. Nevertheless, the effectiveness of the Fenton process may be restricted by variables such as the accessibility of iron ions, the pace at which H_2O_2 decomposes, and the creation of sludge.

9.2.2.4   The Combined Effects of Hydrodynamic Cavitation and the Fenton Process:

- The combination of HC and the Fenton process utilizes the benefits of both techniques, resulting in increased degradation of pollutants and higher efficiency in treatment. The primary synergistic effects encompass:
- The extreme shear pressures and localized high temperatures generated during cavitation can expedite the decomposition of H_2O_2, resulting in the fast production of $^{\bullet}OH$. This increases the overall ability of the Fenton process to undergo oxidation.
- Cavitation enhances mass transfer by disrupting the boundary layers surrounding bubbles and contaminants, hence enhancing the mixing and interaction between pollutants and reactive species. This improves the rates at which mass is transferred, making the breakdown of pollutants more efficient.
- Cavitation can mechanically damage pollutant structures by shear pressures and micro-jets. This process breaks down complex organic compounds and disrupts microbial cell walls, increasing the susceptibility of pollutants to oxidation by $^{\bullet}OH$.
- The synergistic effect of cavitation and the Fenton process facilitates a thorough oxidation of contaminants, hence minimizing sludge formation and the generation of secondary pollutants. This reduces the necessity for additional processing and disposal of sludge.

Case Study 1

Intensification of textile wastewater treatment by H_2O_2, Fenton, and nanofiltration with HC

This study presents the enhancement of wastewater treatment by the synergistic use of HC, Fenton reagent, and nanofiltration method.

In this experiment, an HC reactor of the orifice type was utilized, as depicted in Figure 9.2 of the test section. The compound $FeSO_4 \cdot 7H_2O$ was employed as a Fenton reagent. The initial concentration of total organic carbon (TOC) in the wastewater effluent from textile dyeing was around 1,430 parts per million (ppm).

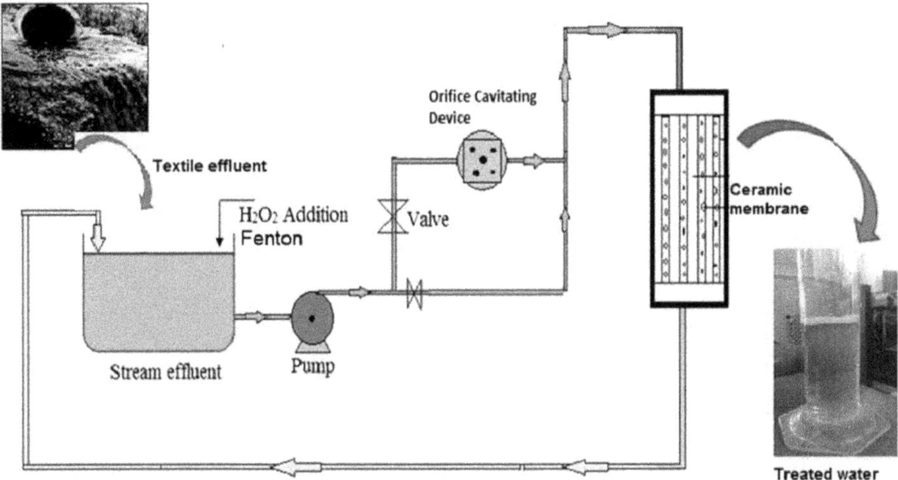

FIGURE 9.2 Experimental setup for combined Fenton, H_2O_2, and hydrodynamic cavitation treatment of effluent.[2]

The administration of H_2O_2 at a specific dosage generates the auxiliary •OH within the cavitation zone, thereby enhancing the effectiveness of these •OH. Consequently, the rate at which pollutants are eliminated is extraordinarily high, in comparison with specific AOPs like H_2O_2 and HC alone. H_2O_2 is commonly recognized as a widely utilized oxidizing agent, possessing an oxidation potential of 1.78 eV. It is frequently employed for the purpose of breaking down pollutants in wastewater. As shown in the following reactions, the combination of HC, Fenton reagent, and nanofiltration resulted in a TOC removal rate of around 59% (Figure 9.3).

$$Fe^{2+} + H_2O_2 \rightarrow Fe^{3+} + •OH + OH^- \tag{9.12}$$

$$Fe^{3+} + H_2O_2 +))) \rightarrow Fe^{2+} + HO_2^• + H^+ \tag{9.13}$$

$$Fe^{3+} + H_2O_2 \rightarrow Fe\text{-}OOH_2 + H^+ \tag{9.14}$$

$$Fe\text{-}OOH_2 +)))) \rightarrow Fe^{2+} + HO_2^• \tag{9.15}$$

Decomposition from HC:

$$H_2O_2 +))) \rightarrow + •OH + •OH \tag{9.16}$$

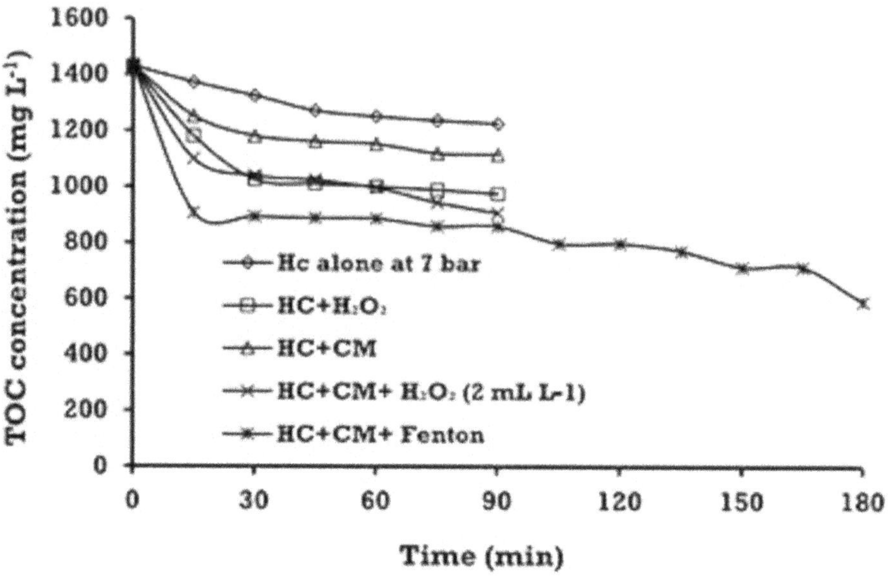

FIGURE 9.3 TOC removal analysis in a combination of H_2O_2, Fenton, and nanofiltration (CM-ceramic membrane) with hydrodynamic cavitation (HC).[2]

9.3 SONO-PHOTOCATALYTIC INACTIVATION OF BACTERIA AND VIRUSES

The presence of harmful microorganisms, such as bacteria and viruses, in water represents a substantial global risk to human health. Conventional disinfection techniques, such chlorination and ultraviolet (UV) irradiation, have constraints in completely deactivating all categories of microorganisms and may occasionally result in the creation of detrimental disinfection byproducts. Advanced disinfection technologies, which integrate various treatment approaches, present a promising alternative to improve the efficiency of microbial inactivation. A novel strategy that can be used is the application of nanofluid-based sono-photocatalytic therapy, which combines the effects of sonication (ultrasound), photocatalysis, and nanotechnology in a synergistic manner.

Sono-photocatalysis is a synergistic method that merges the processes of sonication with photocatalysis to break down contaminants and render microorganisms inactive. Every element of this procedure fulfills a distinct function:

Sonication

- Utilizes ultrasonic waves, often within the frequency range of 20 kHz–1 MHz.
- Induces acoustic cavitation, a phenomenon characterized by the creation, enlargement, and forceful collapse of tiny bubbles, resulting in highly concentrated circumstances such as elevated temperatures and pressures.
- The implosion of cavitation bubbles results in the creation of shock waves and shear stresses, which intensify the transfer of mass, break cell walls, and generate ROS such as •OH.

Photocatalysis:

- Photocatalysis is the process of using light energy, usually UV or visible light, to activate a photocatalyst such as TiO_2.
- Upon exposure to radiation, the photocatalyst initiates the formation of electron-hole pairs, which in turn can generate ROS, such as OH and superoxide anions ($O_2 \bullet^-$).
- These ROS have the ability to oxidize and decompose organic contaminants, as well as render microorganisms inactive by causing damage to their cellular structures and genetic material.

The Role of Nanofluids:

- Nanofluids are precisely designed mixtures of nanoparticles with a base fluid, such as water. Nanofluids have distinct features that make them highly effective in improving the efficacy of sono-photocatalytic applications.
- Nanoparticles possess an increased surface area to volume ratio, which results in a greater number of active sites for photocatalytic processes.
- *Enhanced dispersion*: Nanoparticles can be evenly distributed in the fluid, guaranteeing equal and efficient treatment across the liquid medium.
- Nanoparticles can enhance the production and dispersion of ROS, hence enhancing the process of breaking down pollutants through oxidation.
- Nanoparticles possess the ability to absorb a wider range of light, hence improving their photocatalytic activity under different lighting situations.

Microbial Inactivation Mechanisms:
The synergistic impact of sonication and photocatalysis in the presence of nanofluids results in heightened microbial inactivation through multiple mechanisms (Figure 9.4):

- *Cavitation-induced physical disruption*: The implosion of cavitation bubbles created by ultrasound generates shock waves and shear stresses, which directly and forcefully damage the cell walls and membranes of bacteria and viruses.
- *ROS generation*: Both sonication and photocatalysis produce ROS, including $\bullet OH$ and $O_2 \bullet^-$. ROS have the ability to oxidize various biological components such as lipids, proteins, and nucleic acids, resulting in the inactivation of microorganisms.
- Sonication enhances the process of transferring mass by improving the mixing and dispersion of nanoparticles and ROS in the fluid. This ensures that microorganisms come into contact with the nanoparticles and ROS in a consistent manner.
- Photocatalytic oxidation occurs when the photocatalyst nanoparticles create electron-hole pairs under light irradiation, resulting in the production of extra ROS. The synergistic impact of light and ultrasound amplifies the overall oxidative capability of the therapeutic procedure.

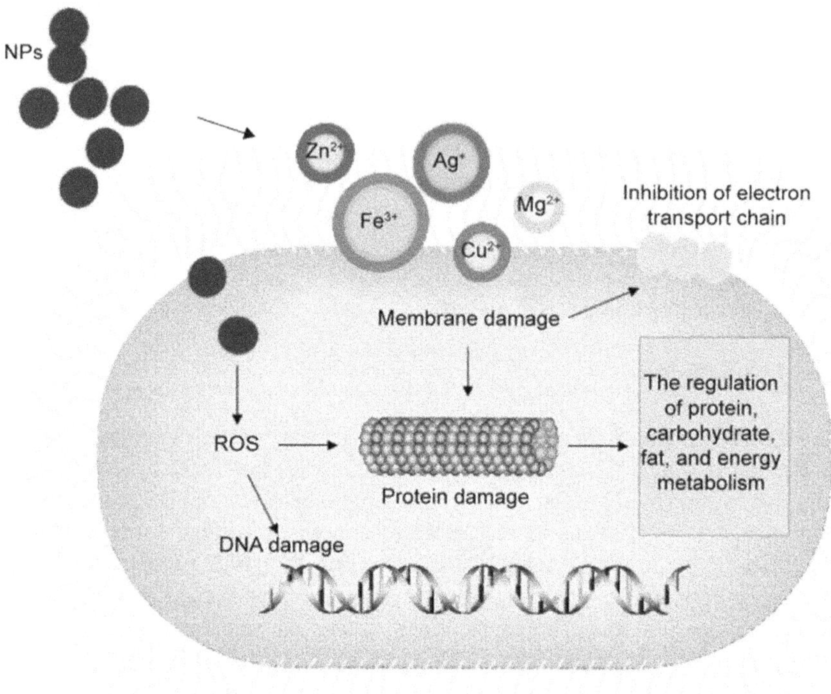

FIGURE 9.4   Inactivation mechanism of bacterial cells using nanoparticles.[3]

9.3.1 INTRODUCTION TO ANTIBACTERIAL AGENTS

There are a variety of industries that can benefit from the use of antibacterial agents. These industries include the environment, food, synthetic fabrics, packaging, healthcare, medical care, as well as building and decoration. There are two basic categories that can be used to classify them: organic and inorganic material. When compared to inorganic antibacterial agents, organic antibacterial materials are typically less stable, particularly when subjected to high temperatures and/or pressures.

The fact that certain inorganic materials, such as metal oxides like TiO_2, zinc oxide (ZnO), magnesium oxide (MgO), and calcium oxide (CaO), are not only stable under harsh process conditions but are also usually considered to be safe materials for both humans and animals makes them of particular interest. In contrast to TiO_2 and ZnO, which have been utilized widely in the development of personal care products, MgO and CaO are minerals that are necessary for human health.

Case Study 1

ZNO NANOFLUID AS AN ANTIBACTERIAL AGENT

ZnO nanofluids, consisting of ZnO nanoparticles dispersed in a base fluid, have attracted considerable interest due to their antibacterial characteristics. The strong antibacterial activity of ZnO nanofluids can be attributed to various inherent qualities of ZnO nanoparticles, such as their distinctive physicochemical characteristics, capacity to produce ROS, and interactions with microbial cells. This study examines the elements that lead to the increased antibacterial effectiveness of ZnO nanofluids.

PROPERTIES OF ZNO NANOPARTICLES

High surface area:

- ZnO nanoparticles exhibit a significant surface area to volume ratio, which results in a substantial active surface area for engaging with bacterial cells. This increases the likelihood of contact and effectiveness of antibacterial effects.

Photocatalytic activity:

- ZnO is widely recognized as a highly effective photocatalyst. ZnO nanoparticles have the ability to produce electron-hole pairs when exposed to UV or visible light.
- This process results in the generation of ROS such as $\bullet OH$, $O_2\bullet^-$, and H_2O_2.
- These ROS possess a high level of reactivity and have the potential to cause harm to the cellular components of bacteria.

Characteristics of semiconductors:

- ZnO is classified as a wide-bandgap semiconductor, indicating its ability to capture light energy and transform it into chemical energy.
- This property enables ZnO to facilitate redox processes that can be detrimental to bacteria.

MECHANISMS OF ANTIBACTERIAL ACTIVITY

Production of ROS:

- The principal method by which ZnO nanoparticles demonstrate antibacterial action is through the production of ROS.
- ZnO nanoparticles, when illuminated, undergo photon absorption and subsequently produce electron-hole pairs.
- Electrons and holes can engage in redox processes, generating ROS that have the potential to harm bacterial cell walls, membranes, proteins, and DNA as shown in Figure 9.5.

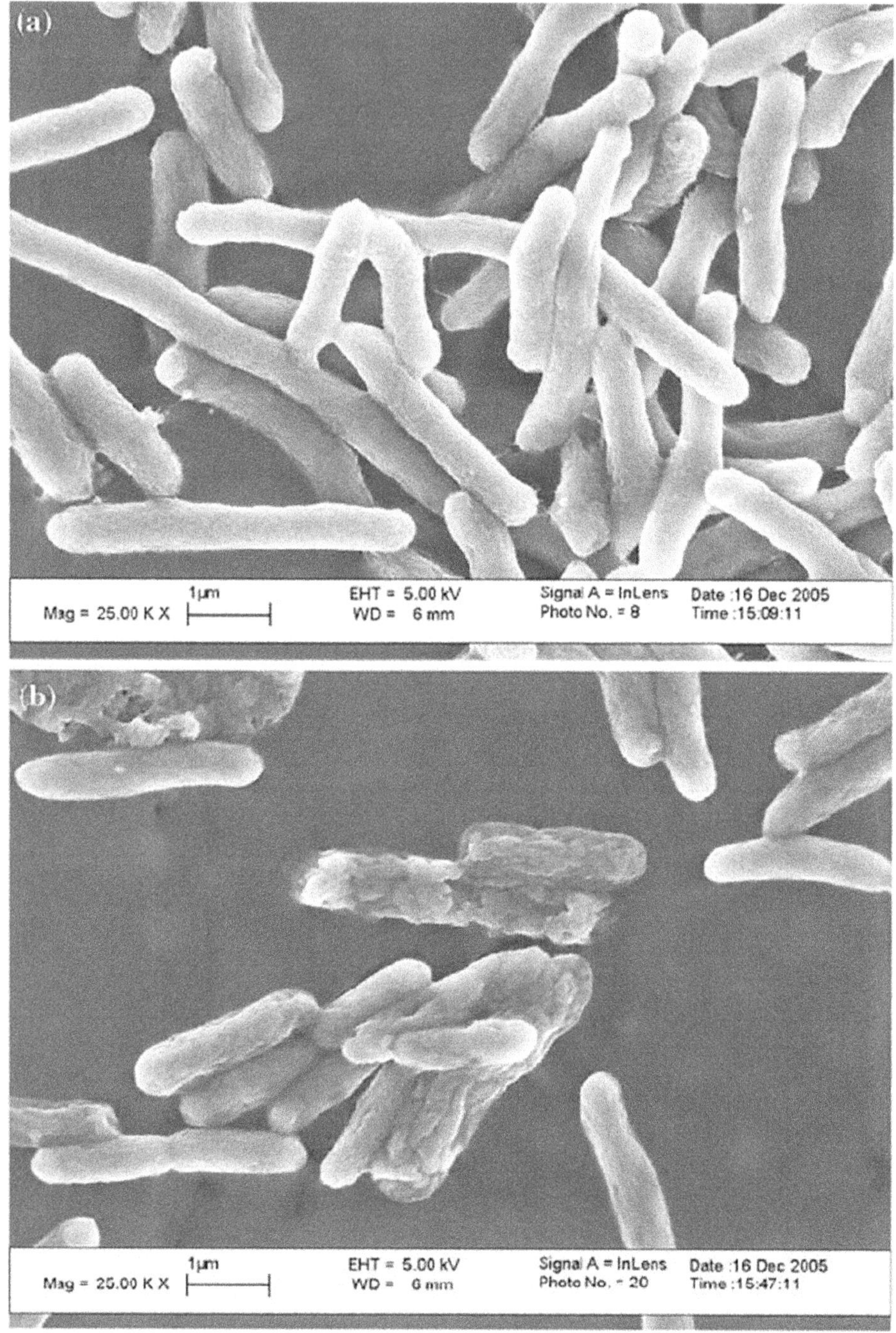

FIGURE 9.5 Scanning electron microscopy images of *Escherichia coli* bacteria: (a) before and (b) after treatment with 0.2% ZnO nanofluids for 5 h.[4]

$$ZnO + h\upsilon \rightarrow ZnO\left(e^- + h^+\right) \tag{9.17}$$

$$e^- + O_2 \rightarrow O_2^{\bullet-} \tag{9.18}$$

$$h^+ + H_2O \rightarrow {}^{\bullet}OH + H^+ \tag{9.19}$$

Direct contact with bacterial cells:

- Electrostatic interactions enable the attachment of ZnO nanoparticles to the bacterial cell wall. Direct contact can result in the mechanical breakage of the cell wall and membrane, leading to the release of intracellular contents and ultimately resulting in cell death.

Release of Zn^{2+} ions:

- In aqueous conditions, ZnO nanoparticles have the ability to release zinc ions (Zn^{2+}). Zn^{2+} with a charge of 2+ are capable of entering bacterial cells and disrupting crucial biological functions, such as enzyme activity and protein synthesis.
- The bactericidal effect of Zn^{2+} ions introduces an additional level of antibacterial activity.
- ZnO can be represented as the dissociation of Zn^{2+} and oxide ions (O_2^-). Therefore, the chemical equation for the dissociation of ZnO is,

$$ZnO \rightarrow Zn^{+2} + O^{2-} \tag{9.20}$$

Photothermal effects:

- ZnO nanoparticles have the ability to transform light energy into heat when exposed to light, resulting in localized hyperthermia.
- Elevated temperatures can induce thermal harm to bacterial cells, hence augmenting the antibacterial effectiveness of ZnO nanofluids.

FACTORS ENHANCING THE ANTIBACTERIAL ACTIVITY OF ZnO NANOFLUIDS

Particle size and shape:

- Smaller ZnO nanoparticles exhibit a greater surface area and are capable of producing a higher amount of ROS per unit mass when compared to bigger particles.
- Moreover, specific geometries, such as rods and needles, can exhibit increased activity as a result of their enhanced capacity to efficiently penetrate bacterial cell walls.

Nanoparticle concentration:

- Increased quantities of ZnO nanoparticles in the nanofluid heighten the probability of contact with bacterial cells and augment the overall antibacterial efficacy.

Light irradiation:

- The presence of UV or visible light greatly increases the effectiveness of ZnO nanofluids in killing bacteria. This is because the light activates photocatalytic activities, which then generate ROS.

Stability and dispersion:

- The ZnO nanoparticles in the nanofluid are stable and evenly distributed, which guarantees a constant antibacterial effect in the entire treated medium. Optimally distributed nanoparticles maximize the available surface area for interactions with bacteria.

APPLICATIONS OF ZNO NANOFLUIDS IN ANTIBACTERIAL TREATMENTS

Water disinfection: ZnO nanofluids have the potential to be employed in water treatment systems for the purpose of disinfecting drinking water and wastewater, thereby guaranteeing the elimination of harmful microorganisms.
Coatings applied to the surface:

- ZnO nanofluids have the potential to serve as effective antibacterial coatings on surfaces in medical facilities, food processing factories, and public areas, effectively inhibiting the growth and spread of microorganisms.

Textiles and personal protective equipment (PPE):

- The addition of ZnO nanofluids to textiles and PPE can provide antibacterial characteristics, hence decreasing the likelihood of infection transmission.

Applications in the pharmaceutical and biomedical fields:

- ZnO nanofluids have the potential to be utilized in pharmaceutical formulations and biomedical devices for the purpose of preventing bacterial infections and improving patient safety.

Case Study 2

SILVER NANOFLUID AS AN ANTIBACTERIAL AGENT

Silver nanofluids, composed of suspended silver nanoparticles (AgNPs) in a base fluid, have been recognized as significant antibacterial agents. The antibacterial qualities of silver have been well established for a long time, and the advancement of nanotechnology has greatly improved its effectiveness. Silver nanofluids possess the advantageous properties of nanoscale silver particles and fluid-based applications, rendering them exceptionally efficient in diverse antibacterial applications.

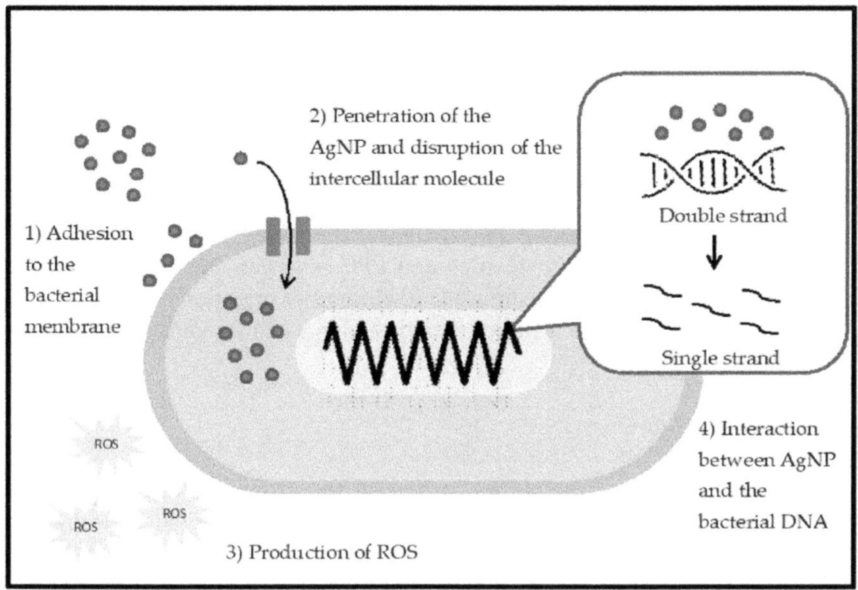

FIGURE 9.6 Inactivation mechanism of bacterial cells using silver nanoparticles.[5]

MECHANISMS OF ANTIBACTERIAL ACTIVITY

Researchers have become interested in AgNPs because of their wide range of antifungal and antibacterial characteristics. There is an urgent need for the development of new antibacterial goods or agents since microbes are becoming resistant to antibiotics through mutation. Furthermore, the relatively low reactivity of AgNPs in comparison with silver ions renders them appropriate for clinical and therapeutic use. AgNPs exhibit four well-documented antibacterial mechanisms:

- Adherence to the microbial surface membrane.
- AgNPs infiltrate the cells and disrupt biomolecules, causing intracellular damage.
- AgNPs create cellular toxicity by producing ROS, which initiate oxidative stress in the cell.
- AgNPs interfere with the signal transduction pathways of the cells. Figure 9.6 illustrates the comprehensive mechanism by which bacteria function.

APPLICATION OF ANTIBACTERIAL TREATMENT IN WATER AND WASTEWATER TREATMENT

The utilization of silver nanofluids is successful in disinfecting drinking water and treating wastewater, leading to the deactivation of a broad spectrum of pathogenic bacteria and guaranteeing the safety of water quality.

Medical devices and implants:

i. AgNPs can be included into medical devices and implants to inhibit bacterial infections and the production of biofilms. Some examples of medical devices are catheters, wound dressings, and surgical equipment.[6]

Coatings applied to the surface:

ii. Silver nanofluids have the potential to serve as antibacterial coatings over a range of surfaces, such as hospital environments, food processing facilities, and public areas, in order to minimize microbial contamination and transmission.

Textiles and PPE:

iii. By integrating AgNPs into textiles and PPE, one can introduce antibacterial characteristics, hence diminishing the likelihood of infection transmission. This is especially beneficial in healthcare environments and for garments that provide protection.

Packaging for food products:

iv. Silver nanofluids have the ability to prevent the growth of microorganisms in food packaging materials, hence increasing the duration for which food products can be stored and maintaining their safety.

Air purification:

v. AgNPs can be employed in air filtration systems to deactivate airborne bacteria and viruses, thereby enhancing indoor air quality and diminishing the likelihood of respiratory diseases.

9.4 CHALLENGES AND LIMITATIONS ASSOCIATED WITH THE UTILIZATION OF NANOFLUIDS IN ANTIBACTERIAL ACTIVITY

Nanofluids, comprising of suspended nanoparticles in a base fluid, possess considerable potential in antibacterial applications owing to their augmented physical, chemical, and biological characteristics. Nevertheless, the utilization of nanofluids in this particular situation is not exempt from obstacles and constraints.[7] This discussion examines the primary obstacles and restrictions related to the utilization of nanofluids for antibacterial objectives.

i Toxicity and environmental impact
- Nanoparticles, although they exhibit efficacy against bacteria, can also present potential hazards to human health. Due to their diminutive size, they can easily pass through biological membranes and accumulate in tissues, which may lead to cytotoxicity, inflammation, and other negative consequences.
- For example: High concentrations of AgNPs have demonstrated toxicity to mammalian cells, which raises issues regarding their safe application in medical and consumer products.

ii. Environmental concerns
- Unintended ecological consequences can arise from the introduction of nanofluids into the ecosystem. Nanoparticles have the potential to accumulate in soil and water, which can have an impact on microbial communities, aquatic species, and plant life.
- The enduring presence and accumulation of nanoparticles in the environment might result in lasting ecological impacts, highlighting the need for meticulous evaluation of their life cycle and proper disposal.

iii. Cost of production
- Producing nanoparticles of excellent quality and stability can incur significant costs, requiring advanced equipment and intricate procedures. This expense is frequently passed on to the final product, rendering nanofluids less economically feasible for extensive utilization.
- The process of transitioning from laboratory to industrial production presents additional difficulties, such as the need to ensure consistent and high-quality nanoparticles.

iv. Stability and dispersion
- Agglomeration refers to the process of particles coming together to form larger clusters, whereas sedimentation refers to the settling of these clusters to the bottom of a liquid or gas.
- Nanoparticles have a tendency to clump together because of the attractive interactions known as van der Waals forces and the random motion known as Brownian motion. This clumping causes the particles to settle down in the liquid they are suspended in. This results in a decrease in the overall surface area and a reduction in the antibacterial effectiveness.
- It is essential to maintain the long-term stability and even distribution of nanoparticles in the fluid in order to achieve consistent antibacterial effectiveness.
- In order to inhibit the formation of clumps, nanofluids are frequently supplemented with surfactants or stabilizers. Nevertheless, these additions may occasionally disrupt the antibacterial characteristics or present more issues regarding toxicity.

v Regulatory and safety concerns
- The production, characterization, and testing of nanofluids currently lack defined techniques. This poses a difficulty in comparing results across several research and ensuring consistent product quality.
- The regulatory frameworks for nanomaterials are still in a state of development, and there is frequent ambiguity around the approval procedure for novel goods that utilize nanofluids.

vi. Safety assessments
- Thorough safety evaluations are necessary to comprehend the potential health hazards linked to nanofluids. This entails examining the toxicity and pharmacokinetics and enduring the impacts of these substances on human health and the environment.
- Securing adherence to regulations and acquiring authorizations from health and environmental agencies can be a laborious and intricate procedure.

vii. Possibility of bacterial resistance development
  - While the mechanisms of nanoparticles decrease the probability of bacterial resistance in comparison to conventional antibiotics, there remains a possibility that bacteria may evolve resistance over time.
  - Prolonged exposure to low levels of nanoparticles may lead to the development of resistant strains, requiring continuous monitoring and adaptive management measures.

viii. Technical challenges in integrating with existing systems
  - Integrating nanofluids into current antibiotic treatment systems and procedures can pose technical difficulties. To prevent any disruptions, it is crucial to ensure compatibility with existing infrastructure and technology.
  - Further innovation and technical solutions are necessary to develop cost-effective and scalable methods for incorporating nanofluids into many applications, including water purification, medical devices, and surface coatings.

ix. Public perception and acceptance
  - The acceptance and adoption of nanofluids can be influenced by the way the general public perceives nanotechnology. Resistance against their use can arise due to concerns regarding safety, potential health risks, and the environmental impact they may have.
  - To address public concerns and foster trust in nanofluid technologies, it is crucial to engage in effective communication and education regarding their advantages and potential risks.

9.5   SUMMARY

Nanofluids have great potential in antibacterial applications because of their improved characteristics and methods of action. Nevertheless, there are certain obstacles and restrictions that must be resolved in order to completely achieve their maximum capabilities. These considerations encompass toxicity, environmental effects, production costs, stability and dispersion, regulatory and safety issues, potential bacterial resistance, technical integration, and public perception. To tackle these issues, it is necessary to conduct research that combines many fields of study, establish strong and comprehensive regulations, and develop communication methods that are efficient in order to guarantee the secure, efficient, and long-lasting utilization of nanofluids in the fight against microbial contamination and the protection of public health.

REFERENCES

1. Thanekar, P., & Parag, G. (2018). Application of HC reactors for treatment of wastewater containing organic pollutants: intensification using hybrid approaches. *Fluids* 3(4), 98. https://doi.org/10.3390/fluids3040098
2. Bethi, B., Sonawane, S. H., Bhanvase, B. A., & Sonawane, S. S. (2021). Textile industry wastewater treatment by cavitation combined with fenton and ceramic nanofiltration membrane. *Chemical Engineering and Processing-Process Intensification, 168*, 108540.

3. Wang, L., Hu, C., & Shao, L. (2017). The antimicrobial activity of nanoparticles: present situation and prospects for the future. *International Journal of Nanomedicine*, *12*, 1227–1249. https://doi.org/10.2147/IJN.S121956.

4. Zhang, L., Jiang, Y., Ding, Y., Povey, M., & York, D. (2007). Investigation into the antibacterial behaviour of suspensions of ZnO nanoparticles (ZnO nanofluids). *Journal of Nanoparticle Research*, *9*, 479–489.

5. Salleh, A., Naomi, R., Utami, N. D., Mohammad, A. W., Mahmoudi, E., Mustafa, N., & Fauzi, M. B. (2020). The potential of silver nanoparticles for antiviral and antibacterial applications: a mechanism of action. *Nanomaterials*, *10*(8), 1566. https://doi.org/10.3390/nano10081566

6. Sonawane, S. S., Khedkar, R. S., & Wasewar, K. L. (2013). Study on concentric tube heat exchanger heat transfer performance using Al_2O_3–water based nanofluids. *International communications in heat and mass transfer*, *49*, 60–68.

7. Khedkar, R. S., Sonawane, S. S., & Wasewar, K. L. (2012). Influence of CuO nanoparticles in enhancing the thermal conductivity of water and monoethylene glycol based nanofluids. *International Communications in Heat and Mass Transfer*, *39*(5), 665–669.

10 Research Progress and Challenges

10.1 RESEARCH PROGRESS AND CHALLENGES OF NANOFLUIDS IN TERMS OF COST ESTIMATION ANALYSIS

Nanofluids have attracted considerable attention for their ability to improve heat and mass transfer, owing to their improved properties in comparison with traditional fluids. Nevertheless, the economic viability of these new materials continues to be a significant challenge for their widespread implementation. This analysis provides a thorough examination of the several elements involved in estimating the costs of nanofluids. It covers production costs, material costs, scaling challenges, life cycle costs, and economic factors relevant to different applications.

10.1.1 Costs of Nanoparticle Synthesis

The production of nanoparticles is a major factor in the total expense of nanofluids. There are multiple techniques available for producing nanoparticles, each with its own prices and efficiencies.

Chemical vapor deposition (CVD)

- Chemical vapor deposition (CVD) is a commonly employed technique for manufacturing nanoparticles with a high level of purity. This technique entails the chemical transformation of gaseous precursors into solid nanoparticles on a substrate. Although CVD has the capability to generate nanoparticles of excellent quality, the approach is costly due to the need for high temperatures and complex equipment.
- Various factors, including precursor materials, energy consumption, and equipment maintenance, affect the cost of CVD. The significant capital and operational expenses restrict the economic feasibility of CVD for large-scale manufacturing.

The sol-gel process

- The sol-gel process is the transformation of a liquid solution into a solid gel phase. This approach is beneficial due to its capacity to generate consistent nanoparticles at relatively low temperatures. Nevertheless, the expenses can be amplified due to the cost of precursors and solvents, as well as the requirement for post-synthesis heat treatment.

DOI: 10.1201/9781003594949-10

- While the sol-gel technique is cost-effective compared to CVD, it still poses challenges in terms of maintaining consistent quality and scaling up production. These challenges have an impact on the overall cost-effectiveness of this method.

Laser ablation

- Laser ablation is a process that utilizes intense laser pulses to evaporate material from a solid object, resulting in the creation of nanoparticles. This approach provides accurate manipulation of particle size and content. Nevertheless, the considerable energy consumption and requirement for specialist equipment render laser ablation a costly alternative.
- The scalability of laser ablation is constrained due to the need for substantial expenditures in high-power laser systems and comprehensive safety precautions in order to achieve large-scale output.

Alternative approaches

- Hydrothermal synthesis, microemulsion techniques, and flame spray pyrolysis are further methods that enhance the variety of techniques used for producing nanoparticles. Every approach possesses an own cost framework determined by factors such as raw materials, energy demands, and possibilities for scalability.

10.1.2 Costs Associated with Materials

The selection of materials for nanoparticles and base fluids has a substantial influence on the total cost of nanofluids.

Primary resources

- The cost of raw materials exhibits significant variation based on the specific type of nanoparticles employed. Metals such as silver and gold possess a high price tag, resulting in the elevated cost of their nanoparticles. On the other hand, oxides like aluminum oxide (Al_2O_3) and silicon dioxide (SiO_2) are less expensive.
- Carbon-based nanoparticles, such as carbon nanotubes (CNTs) and graphene, are costly because their production procedures are intricate.

Base fluids

- The cost is also influenced by the choice of base fluid in which the nanoparticles are disseminated. Typical base fluids consist of water, ethylene glycol, and other types of oils. Although water is the most cost-effective choice, certain specific uses may necessitate the use of pricier base fluids in order to achieve the best possible performance.

Additives and stabilizers

- In order to attain consistent dispersals, a variety of substances, including surfactants and polymers, are employed as additives and stabilizers. These compounds contribute to the total expense, particularly when significant amounts are required to sustain the stability of nanoparticles over a period of time.

10.1.3 EXPENSES RELATED TO DISPERSION AND STABILIZATION

It is essential to maintain the steady dispersion of nanoparticles in the base fluid in order to preserve the improved thermal characteristics of nanofluids.

Methods of dispersion

- Common methods for dispersing nanoparticles include mechanical agitation, ultrasonication, and high-shear mixing. Every approach has expenses associated with equipment, energy usage, and the duration of the procedure.
- Ultrasonication necessitates the use of ultrasonic processors, which can be costly and use a significant amount of energy, particularly when dealing with big quantities.

Stabilizing agents

- Utilizing stabilizing chemicals is crucial in order to prevent the clumping and settling of nanoparticles. Surfactants, polymers, and surface functionalization processes are utilized to improve stability.
- The price of these stabilizers is contingent upon the specific type and quantity employed. Stabilizers that guarantee long-term stability at a high level of performance can be quite expensive.

10.1.4 SCALABILITY AND PRODUCTION COSTS

Transitioning the manufacturing of nanofluids from a laboratory setting to an industrial scale poses substantial financial obstacles.

Equipment and infrastructure

Extensive production necessitates significant expenditures in specialized machinery and infrastructure. This encompasses reactors designed for the creation of nanoparticles, mixing equipment utilized for dispersion purposes, and storage facilities specifically designed for nanofluids. In addition to expenses related to expansion, the cost of scaling up also encompasses alterations to current facilities or the construction of new manufacturing plants.

Quality control

- It is crucial to maintain a constant level of quality when doing production on a wide scale. Robust quality control methods are necessary, which involve real-time monitoring, analytical testing, and strict adherence to demanding standards.

- Integrating quality control systems increases operational expenses due to the requirement of advanced equipment and highly trained staff.

Regulatory compliance

- Industrial-scale production necessitates strict adherence to regulatory requirements and safety norms. These encompass environmental restrictions, health and safety standards, and product certifications.
- Adhering to these regulatory requirements results in extra expenses for conducting tests, creating documentation, and managing compliance.

Life cycle costs

- An all-encompassing cost calculation should take into account the complete life cycle of nanofluids, spanning from their creation to their disposal.

Costs associated with the production process

- The initial production costs include the expenses associated with acquiring raw materials, the process of synthesizing the product, dispersing it, and stabilizing it. The costs fluctuate depending on the selected techniques and materials.

Operational costs

- Utilizing nanofluids in different applications involves expenses associated with the upkeep, supervision, and regular replacement of fluids to uphold function.
- The durability of nanofluids over an extended period of time has a direct impact on their overall operational expenses. Regular replacement or re-stabilization leads to higher costs.

Disposal and environmental costs

- Compliance with environmental rules is necessary for the proper disposal of nanofluids in order to avoid any ecological damage. This entails the process of treating and securely disposing of utilized nanofluids, which can incur significant expenses.

10.1.5 LCA Costs

Life cycle assessment (LCA) assesses the ecological consequences of nanofluids at every stage of their existence. Implementing sustainable production practices and providing recycling possibilities can reduce disposal costs and improve economic viability.

Application-specific economic considerations
The economic viability of nanofluids differs among various applications, contingent upon the value they contribute and the industry's sensitivity to costs.
Applications of significant value

- Nanofluids can offer significant performance improvements in applications that require high value, such as electronic cooling and pharmaceuticals, which can justify the associated higher prices. The enhanced thermal conductivity and efficiency of heat transfer result in superior performance and dependability, compensating for the original expenditure.
- In the electronics sector, the exceptional heat dissipation abilities of nanofluids can effectively prevent excessive heating and prolong the durability of components, resulting in notable cost advantages.

Cost-sensitive applications

- In industries with a focus on cost, such as HVAC (heating, ventilation, and air conditioning) and automotive cooling, the use of nanofluids is hindered by the requirement for affordable solutions. The financial benefits resulting from enhanced energy efficiency and decreased maintenance expenses must surpass the original greater investment.
- Attaining an equilibrium between improving performance and ensuring cost-effectiveness is essential for achieving universal acceptance in these industries.

Industrial processes

- Integrating nanofluids into industrial processes, such as chemical manufacture and power generation, poses distinctive difficulties. It is necessary to take into account the expenses associated with modifying current systems and assuring their compatibility with nanofluid technology.
- Nanofluids can facilitate heat transmission in industrial processes, resulting in increased efficiency and decreased energy usage. Nevertheless, the economic viability is contingent upon the extent of implementation and the particular demands of each procedure.

10.1.6 AREAS FOR FURTHER EXPLORATION AND REQUIREMENTS FOR FURTHER RESEARCH

In order to enhance the economic viability of nanofluids, future studies should concentrate on four crucial domains.
Efficient synthesis techniques at a reasonable cost
It is crucial to develop novel synthesis techniques that can lower manufacturing expenses without compromising on the quality. Research should investigate alternate sources of raw materials and employ creative methods of synthesis in order to accomplish this objective.

Scalability solutions
To overcome scalability issues, it is necessary to make breakthroughs in manufacturing technologies and optimize processes. The collaboration between academia and industry can expedite the creation of scalable production processes.

Stability and longevity
Improving the long-term durability of nanofluids can save operational expenses by reducing the frequency of replacement or re-stabilization. The research should prioritize the identification of efficient stabilizers and dispersion strategies.

Environmental impact and sustainability
It is essential to conduct thorough LCA in order to analyze the environmental impact of nanofluids and determine sustainable solutions for their manufacture and disposal. This involves investigating recycling possibilities and minimizing the environmental impact of nanofluid technologies.

Application-specific research
Customizing nanofluid compositions for particular uses might enhance efficiency and cost-efficiency. Engaging in collaborative research with industry partners can result in tailored solutions that effectively address the distinct needs of various sectors.

Analysis of economic models and cost-benefit assessment
Creating intricate economic models and performing comprehensive cost-benefit evaluations for different uses can offer significant perspectives on the financial feasibility of nanofluids. These models should take into account elements such as the initial investment, operational cost reductions, and long-term advantages.

10.2 INDUSTRIAL OPERATION AND SCALE-UP CHALLENGES FOR NANOFLUID APPLICATIONS

The utilization of nanofluids and hybrid nanofluids in industrial processes shows great potential for a wide range of applications. However, there are certain obstacles that need to be overcome in order to ensure their successful implementation. Nanofluids are colloidal suspensions of nanoparticles in a base fluid that have been engineered. Hybrid nanofluids, on the other hand, are a combination of different types of nanoparticles or have additives added to them. Both nanofluids and hybrid nanofluids provide improved thermal conductivity, enhanced heat transfer rates, and other desirable properties when compared to traditional heat transfer fluids. Nevertheless, the implementation of these technologies in industrial settings has challenges pertaining to manufacturing, stability, cost, safety, and environmental considerations. This thorough investigation examines the obstacles and possible remedies in great detail.

10.2.1 CHALLENGES WITH NANOFLUID PRODUCTION

10.2.1.1 Dispersing Nanoparticles

One of the primary challenges in industrial processes that use nanofluids is the attainment of homogeneous dispersion of nanoparticles within the base fluid. Agglomeration, settling, and inadequate dispersion may arise as a result of van der Waals forces, electrostatic interactions, and constraints imposed by Brownian motion. As a result, there are variations in the characteristics and diminished efficiency. It is crucial to develop strong dispersion techniques, such as ultrasonication, mechanical stirring, or surface modification of nanoparticles, in order to ensure stability and optimize performance.

10.2.1.2 Scalability

Scaling up the manufacturing of nanofluids from the laboratory to an industrial level presents considerable difficulties. Traditional methods of synthesis, such as chemical precipitation or sol-gel processes, may not be easily expandable or economically feasible. To achieve larger production volumes while ensuring consistent nanoparticle size distribution, purity, and stability, it is necessary to employ creative engineering techniques and optimize various process parameters.

10.2.2 PROBLEMS RELATED TO STABILITY

10.2.2.1 Sedimentation and Agglomeration

The presence of nanoparticles in nanofluids has a tendency to gradually settle down, resulting in sedimentation and stratification, particularly in situations where there is no movement or very low fluid flow. The process of agglomeration worsens this problem by decreasing the actual surface area and the improvement in heat conductivity. To address stability difficulties, it is necessary to optimize the particle size, shape, and surface chemistry in order to reduce aggregation. Additionally, stabilizing chemicals or surfactants should be used.

10.2.2.2 Thermal Decomposition

High temperatures and extended thermal cycling can lead to the deterioration of nanofluid stability. The surfactants or organic additions may undergo heat breakdown, resulting in phase separation or alterations in viscosity and thermal conductivity. It is crucial for industrial applications to develop formulations that are thermally stable and to investigate new stabilizers that can withstand heat degradation.

10.2.3 COST ANALYSIS

10.2.3.1 Cost of Nanoparticles

The exorbitant expense of nanoparticles, especially those possessing distinctive characteristics or customized capabilities, poses a substantial obstacle to their extensive use in industrial operations. Mass production methods and the benefits of producing on a large scale can lower expenses, but it is essential to develop inventive approaches for producing or reusing nanoparticles in a cost-efficient manner in order to improve affordability and competitiveness.

10.2.3.2 Costs of Processing

In addition to the expenses related to nanoparticles, the supplementary costs linked to the synthesis, dispersion, and quality control of nanofluids also influence the total cost-effectiveness. Efficiently optimizing manufacturing processes, reducing energy usage, and simplifying production workflows are crucial for reducing processing expenses and enhancing the economic viability of large-scale industrial operations.

10.2.4 Safety and Environmental Considerations

10.2.4.1 Hazards to Health and Safety

Workers may face health and safety hazards when handling, synthesizing, and disposing of nanofluids containing nanoparticles. Exposure to nanoparticles by inhalation or contact with the skin may result in respiratory issues, skin irritation, or enduring health consequences. It is crucial to implement strict safety regulations, engineering controls, and protective measures, as well as conduct extensive risk assessments, in order to guarantee occupational health and reduce potential dangers.

10.2.4.2 Environmental Impact

An additional area of concern is the environmental ramifications associated with the manufacture, utilization, and disposal of nanofluids. The release or discharge of nanofluids into ecosystems can result in the introduction of nanoparticles into water bodies or soil, which might have ecological consequences and the ability to accumulate in the food chain. It is crucial to engage in the development of environmentally friendly nanofluid formulations, the investigation of biodegradable nanoparticles, and the implementation of efficient waste management systems in order to decrease environmental impacts and advance sustainability.

10.2.5 Enhancing Performance

10.2.5.1 Enhancing Heat Transfer

Although nanofluids exhibit higher thermal conductivity than traditional heat transfer fluids, the primary goal in industrial operations is still to maximize heat transfer enhancement. By optimizing the concentration, size distribution, and dispersion quality of nanoparticles, as well as making improvements to the system design, implementing flow enhancement techniques, and optimizing the heat exchanger, we may improve heat transfer performance and increase energy efficiency.

10.2.5.2 Operational Stability

Nanofluid stability can be compromised by the harsh working conditions commonly seen in industrial processes, such as high temperatures, pressures, and mechanical stresses. To ensure strong stability in various operating conditions, it is necessary to conduct thorough performance testing, verify material compatibility, and carry out dependability studies. It is crucial to create nanofluid formulations that can withstand harsh circumstances in order to provide dependable and durable industrial use.

10.3 LIFE CYCLE ANALYSIS OF NANOFLUIDS

LCA is a methodical strategy used to evaluate the environmental effects linked to a product, process, or service across its complete life cycle, starting with the extraction of raw materials to the final disposal or recycling stage. Performing a Life Cycle

Assessment (LCA) is crucial for comprehending the environmental consequences of different activities and guiding decision-making processes to advance sustainability and optimize resource utilization. This text provides a comprehensive examination of the significance of undertaking LCA, along with a detailed and systematic approach to carrying out an LCA.

10.3.1 Significance of Conducting Life Cycle Analysis

A comprehensive assessment: LCA offers a comprehensive perspective on the environmental consequences linked to a product or process, taking into account all phases of its life cycle. This comprehensive assessment enables stakeholders to identify and prioritize areas with significant environmental impact, examine the trade-offs involved, and formulate effective policies for environmental management and enhancement.

LCA results are essential tools for businesses, governments, and consumers to support decision-making. They help guide product design, process optimization, investment decisions, and policy creation. LCA allows for the measurement and comparison of environmental effects and provides the necessary information for making well-informed decisions that take into account both environmental concerns and economic and social aspects.

Resource efficiency is a key benefit of LCA since it enables the identification of possibilities to reduce waste, minimize contamination, and optimize resource usage over the whole life cycle of products and activities. LCA helps in sustainable resource management and circular economy concepts by optimizing material and energy flows, lowering resource consumption, and reducing environmental impacts.

LCA enables the recognition and evaluation of potential environmental hazards and consequences linked to various phases of a product's life cycle. Through the assessment of variables such as emissions, waste production, and toxicity, LCA aids in the reduction of risks, guaranteeing adherence to regulations and minimizing negative environmental impacts.

LCA enhances openness and accountability through the use of a methodical and uniform framework for evaluating environmental impact. Through the documentation of data sources, assumptions, and methodology, the outcomes of LCA can be thoroughly examined, validated, and effectively conveyed to stakeholders. This practice fosters confidence, credibility, and accountability in decision-making processes.

Continuous improvement: LCA functions as a mechanism for ongoing enhancement and originality, propelling sustainability endeavors, product advancement, and process streamlining. LCA promotes sustainability and encourages firms to continuously learn and innovate by identifying areas for environmental improvement and establishing reduction targets.

10.3.2 Procedure for Conducting Life Cycle Analysis

10.3.2.1 Definition of Objectives and Scope

Specify the goals and limitations of the LCA study, which involve identifying the specific product or system to be evaluated, determining the functional unit of analysis, and selecting the environmental impact categories that are of interest. Define the limits of the system, which should encompass all stages of the life cycle such as raw material extraction, production, usage, and disposal. Also, identify any activities or impacts that are not included in the analysis.[2]

10.3.2.2 Life Cycle Inventory (LCI)

LCI refers to the systematic collection and quantification of inputs, outputs, and environmental impacts associated with a product or process throughout its entire life cycle. Compile a thorough inventory of all inputs (such as materials, energy, and water) and outputs (such as emissions and waste) that are linked to each stage of the life cycle. Gather data on the utilization of resources, release of emissions, and production of trash from original sources, comprehensive literature reviews, databases, and consultations with experts.[3]

10.3.2.3 Life Cycle Impact Assessment (LCIA)

LCIA is a method used to evaluate the potential environmental impacts of a product or process throughout its entire life cycle. Assess the possible environmental effects linked to the LCI data by utilizing impact assessment methodologies and models. Utilize characterization parameters to measure the magnitude of affects in many environmental impact categories, including climate change, resource depletion, acidification, and human toxicity.

10.3.2.4 Explanation

Analyze the results of LCA to identify regions with high environmental impact, determine the main factors contributing to these impacts, and pinpoint opportunities for improvement. Analyze the compromises between several environmental impact categories and take into account the uncertainty and sensitivity of LCA results.

10.3.2.5 Analysis of Improvement

Identify potential areas for environmental improvement and enhancement of sustainability based on the findings of an LCA. Create and implement plans and procedures to minimize the negative effects on the environment, maximize the efficient use of resources, and decrease potential risks at every stage of the product's life cycle.

10.3.2.6 Documentation and Dissemination of Information

Produce a report that clearly and thoroughly outlines the LCA approach, assumptions, data sources, and results in a manner that is easily understood and leaves no room for ambiguity. Effectively convey the findings of an LCA to stakeholders through reports, presentations, and other communication channels. Ensure that the communication is clear, accurate, and relevant to the decision-making processes.

10.3.2.7 Evaluation and Verification

Conduct a thorough examination and verification of the LCA research by subjecting it to peer review, gathering input from stakeholders, and implementing quality assurance procedures. To improve the credibility and reliability of LCA results, it is important to address any uncertainties, restrictions, or discrepancies that are

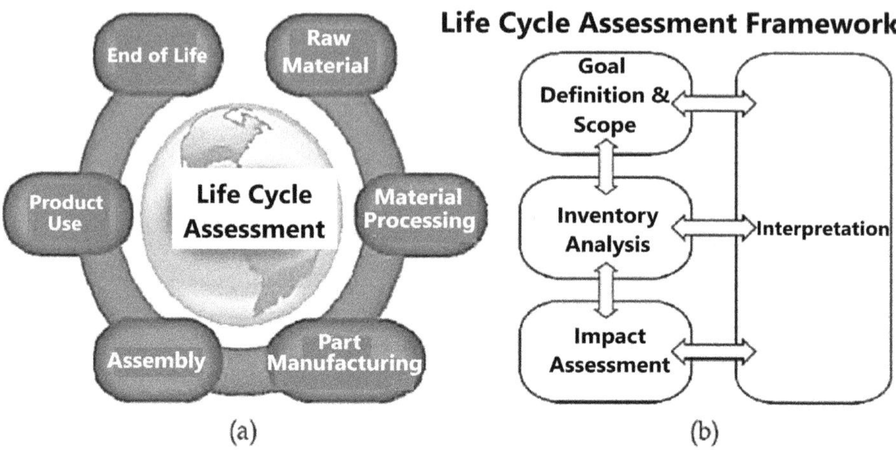

FIGURE 10.1 Illustration depicting the schematic design of (a) cradle to grave life cycle assessment (LCA) and (b) LCA framework applied to nanofluids.[1]

detected throughout the review process. By adhering to these procedures, those with a vested interest can carry out a comprehensive and enlightening examination of the entire lifespan of a product or process. This study yields useful knowledge regarding the ecological consequences and aids in making informed choices, ultimately fostering sustainable progress.[4]

Note

The fundamental principles and methodologies of LCA are applicable to both nanofluids and hybrid nanofluids. However, the evaluation of hybrid systems requires additional considerations regarding the integration of multiple types of nanoparticles, interactions between nanoparticles, and potential synergistic effects on environmental performance.

10.3.3 Case Study

Conduct cradle to grave LCA and compare LCA of ZnO/water nanofluid with CuO-ZnO/water hybrid nanofluid.

The comparison of LCAs between ZnO water nanofluid and CuO-ZnO/water hybrid nanofluid entails evaluating the environmental effects linked to each nanofluid formulation over their complete life cycles (Figure 10.1). Below is a comparative analysis focusing on major elements of their life cycle stages:

- *Extraction of raw materials and production of nanoparticles*

    CuO nanoparticles: The procurement of CuO nanoparticles' raw materials, usually copper ore, entails laborious mining activities and chemical refinement that consume significant amounts of energy. This process is responsible for resource depletion, habitat damage, and the release of contaminants.

ZnO nanoparticles: The acquisition of raw materials for ZnO nanoparticles, including zinc ore, entails mining and processing operations that have environmental consequences. These consequences include energy usage, land disruption, and emissions.

- *Production of nanofluids*

  The production of ZnO nanofluid includes the dispersion of ZnO nanoparticles into a water-based solution, which necessitates the use of energy-intensive dispersion processes and perhaps the addition of chemical additives for stabilization. This process is responsible for energy consumption, chemical utilization, and wastewater production.

  The production of CuO-ZnO hybrid nanofluid entails the amalgamation of CuO and ZnO nanoparticles in a water-based solution, similar to the production process of CuO nanofluid. Furthermore, it is important to take into account the optimization of nanoparticle ratios, dispersion techniques, and the compatibility between CuO and ZnO nanoparticles.

- *Utilization of the Product*

  The CuO nanofluid is utilized in heat transfer systems to augment thermal conductivity and optimize energy efficiency. The environmental effects during usage are mostly linked to energy consumption, system functioning, and maintenance. The CuO-ZnO hybrid nanofluid, like the CuO nanofluid, is utilized to improve heat transmission and has comparable environmental effects when applied. Nevertheless, the collective impacts of CuO and ZnO nanoparticles can potentially affect the efficiency of heat transmission and operational performance.

- *Disposal or recycling at the end of the product's life*

  The disposal of CuO/water nanofluid products at the end of their life cycle has the potential to release CuO nanoparticles into the environment, which could pose dangers to ecosystems and human health. The recycling of CuO nanoparticles is constrained by technological difficulties and economic viability. The CuO-ZnO hybrid nanofluid, when disposed of, can potentially release both CuO and ZnO nanoparticles into the environment, posing environmental hazards. Specialized techniques and considerations are necessary for the recycling or recovery of CuO-ZnO nanoparticles, including nanoparticle separation and purification.

- *Environmental impacts in general*

  The environmental effects of CuO/water nanofluid are mainly determined by the synthesis, dispersion, and utilization of CuO nanoparticles. Important factors to consider include the amount of energy used, the chemicals used, the amount of waste produced, and the potential environmental dangers related to CuO nanoparticles. The environmental effects of CuO-ZnO/water hybrid nanofluid are determined by the manufacturing, dispersion, and utilization of CuO and ZnO nanoparticles. Other factors to take into account include the cumulative impacts of CuO and ZnO nanoparticles on environmental performance, as well as the possibility of synergistic or antagonistic interactions.

To summarize, the comparison of life cycle studies between CuO/water nanofluid and CuO-ZnO/water hybrid nanofluid entails assessing the environmental effects related to the extraction of raw materials, creation of nanoparticles, manufacturing processes, usage of the products, and disposal or recycling at the end of their life cycle. Although both nanofluid formulations have parallels in their life cycle stages, the presence of ZnO nanoparticles in the hybrid nanofluid brings about new factors to consider, such as nanoparticle compatibility, environmental dangers, and overall environmental performance.

10.4 SWOT (STRENGTH, WEAKNESS, OPPORTUNITIES, AND THREATS) OF NANOFLUIDS

10.4.1 STRENGTHS

Nanofluids demonstrate a notable increase in thermal conductivity when compared to conventional heat transfer fluids. This characteristic allows for more effective heat exchange and enhanced energy efficiency in a wide range of industrial applications. Nanofluids and hybrid nanofluids provide a high degree of adaptability in their composition and use, making them suitable for a wide range of industries, including electronics cooling, renewable energy systems, manufacturing processes, and biological applications. Novelty in nanotechnology enables the customization of nanofluid properties, such as particle size, shape, surface chemistry, and dispersion characteristics to meet specific industrial needs and improve performance.

Most importantly, nanofluids improve heat transfer rates, decrease system size and weight, boost equipment reliability, and prolong operating lifetimes, resulting in overall performance optimization in industrial processes.

10.4.2 WEAKNESSES

The exorbitant expenses associated with nanoparticles and manufacturing procedures pose a substantial obstacle to the broad acceptance of nanofluid technologies. This hinders the economic viability and scalability of these technologies, especially in cost-conscious industrial sectors. Nanofluids may encounter stability issues such as particle aggregation, sedimentation, and deterioration over time. These challenges can compromise the performance and reliability of industrial operations, particularly in extreme working circumstances.

The production of nanofluids is a complex process that involves the synthesis of nanoparticles, dispersion, and formulation of the fluids. It requires specialized equipment, experience, and quality control methods. These complexities might make it challenging to scale up production and achieve mass production.

The presence of nanoparticles in nanofluids gives rise to issues regarding health and safety, specifically in relation to occupational exposure, inhalation risks, and environmental effects. As a result, it is necessary to conduct comprehensive risk assessments, ensure compliance with regulations, and implement protective measures in industrial settings.

10.4.3 Opportunities

Ongoing developments in the synthesis of nanoparticles, modification of their surfaces, and techniques for dispersing them present possibilities for enhancing the stability, performance, and cost-effectiveness of nanofluids. This drives innovation and broadens the range of potential applications. The increasing need for energy-efficient solutions, advanced materials, and sustainable technologies has led to a rise in demand for nanofluids in emerging fields such as advanced electronics cooling, renewable energy systems, and biomedical devices. This has resulted in market growth and diversification.

Collaborative partnerships between researchers, manufacturers, policymakers, and industry stakeholders promote the sharing of knowledge, transfer of technology, and development of markets, which speeds up the process of industrializing and commercializing nanofluid innovations. The sustainability focus prioritizes the development of eco-friendly nanofluid formulations, green synthesis processes, and recyclable materials, in line with global sustainability goals and regulatory requirements, to promote environmental responsibility.

10.4.4 Threats

Nanofluid manufacturers have hurdles in complying with evolving regulatory frameworks, safety standards, and environmental restrictions. They must adhere to strict requirements and use risk management procedures to ensure compliance with regulations and gain market access. The market for nanofluid solutions faces significant challenges in terms of market penetration and adoption due to intense competition from alternative heat transfer fluids, conventional technologies, and emerging materials. To stay competitive, it is crucial to employ differentiation, value proposition, and market positioning strategies.

The presence of market volatility, economic swings, and geopolitical issues introduces uncertainty and risk into the nanofluid market, which in turn affects investment decisions, availability of funding, and market demand for nanofluid technology. Public perception has a crucial role in shaping consumer acceptance, regulatory scrutiny, and market demand for nanofluid goods. Misconceptions and concerns about the safety of nanoparticles, their environmental impact, and potential health risks can significantly impact the growth and adoption rates of these products.

10.5 FUTURE PERSPECTIVES OF NANOFLUIDS

Nanofluids, which are combinations of nanoparticles and regular base fluids, have shown great promise in improving heat transfer efficiency in a wide range of applications. Their distinctive thermal properties present encouraging possibilities for surpassing the limitations of conventional heat transfer fluids. The growing demand for more effective heat management systems in industry has led to significant interest in the potential industrialization of nanofluids in the future. This section examines the possible future paths, obstacles, and prospects for the widespread industrial implementation of nanofluids.[5]

i. **Advancements in Nanoparticle Synthesis and Stability**

Improved Synthesis Techniques:

Subsequent investigations are expected to prioritize the advancement of more proficient and economical techniques for synthesizing nanoparticles. Methods such as CVD, sol-gel procedures, and laser ablation are anticipated to be improved, resulting in nanoparticles with accurately regulated dimensions, morphology, and surface characteristics. Advanced synthesis techniques will enable the creation of high-quality nanofluids with exceptional thermal characteristics.

Enhanced Stability:

Ensuring long-term stability is a major obstacle in the commercialization of nanofluids. Future improvements will focus on the development of enhanced dispersants and surfactants to mitigate the issues of agglomeration and sedimentation of nanoparticles. Surface modification approaches, such as polymer coating functionalization or the application of electrostatic and steric stabilization methods, will be vital in improving the stability of nanofluids.

ii. **Tailoring Nanofluids for Specific Applications**

Customized Formulations

The capacity to customize nanofluid compositions to fulfill the specific requirements of various industries will propel their industrialization. In the electronics industry, nanofluids that possess high heat conductivity and low electrical conductivity are favored in order to prevent short circuits. Similarly, the pharmaceutical industry will require biocompatible nanofluids that have precise control over their viscosity and temperature properties. These nanofluids will play a crucial role in applications like targeted drug administration and biomedical imaging.

Multi-functional Nanofluids

Subsequent investigations will focus on the advancement of multi-functional nanofluids that have additional benefits beyond improved thermal conductivity. For instance, magnetic nanofluids, also known as ferrofluids, may be easily controlled by magnetic fields, allowing for accurate regulation of heat transport in a wide range of applications. Moreover, the utilization of nanofluids having antibacterial characteristics holds potential in sectors like food processing and healthcare, serving the purpose of averting contamination and enhancing safety.

iii. **Environmental and Economic Considerations**

Eco-Friendly Nanofluids

The increasing environmental concerns have led to a substantial emphasis on the development of nanofluids that are environmentally friendly. The researchers will study the application of biodegradable nanoparticles and green production techniques in order to reduce the negative effects on the environment. In addition, the management of nanofluid recycling and disposal will be discussed to ensure the implementation of sustainable practices in industrial applications.

Cost-Effectiveness

In order for nanofluids to be widely adopted in industry, they need to be economically feasible. Future endeavors will focus on decreasing the expense of nanoparticle manufacturing and enhancing the scalability of nanofluid synthesis procedures. Effective collaboration among academia, industry, and government entities will be essential for financing research and development projects aimed at reducing costs and establishing nanofluids as a cost-efficient option for diverse applications.

iv. **Integration with Emerging Technologies**

Internet of Things (IoT) and Smart Systems

The incorporation of nanofluids into internet of things (IoT) and intelligent systems would profoundly transform thermal management in several industries. Advanced sensors and control systems have the ability to continuously monitor and regulate the characteristics of nanofluids, resulting in improved heat transfer performance and energy efficiency. IoT-enabled nanofluids in HVAC systems may adapt their thermal conductivity in response to the surrounding temperature, allowing for efficient cooling or heating.

Renewable Energy Systems

Nanofluids have the potential to significantly improve the efficiency of renewable energy sources. Nanofluids with excellent thermal conductivity can enhance the absorption and transmission of solar energy in solar thermal collectors. In geothermal systems, the use of nanofluids can improve the transfer of heat between the Earth and the working fluid, resulting in a higher overall efficiency of the system. Subsequent investigations will prioritize the enhancement of nanofluid compositions for these specific uses and their seamless integration with current renewable energy technology.

v. **Standardization and Regulation**

Establishing Industry Standards

In order to promote the industrialization of nanofluids, it is crucial to create standardized methods for their synthesis, characterization, and use. Adhering to industry standards will guarantee that nanofluids perform consistently and reliably in various applications. The development of these standards will rely heavily on the collaborative efforts of regulatory agencies, academic institutes, and industry stakeholders.

Regulatory Framework

With the increasing use of nanofluids in industrial applications, it is crucial to develop regulatory frameworks to guarantee their safe utilization. The rules encompass protocols for the management, preservation, and elimination of nanofluids, alongside restrictions pertaining to thresholds of exposure and ecological consequences. Future endeavors will prioritize the development of all-encompassing regulatory frameworks that specifically target the distinct characteristics and possible hazards of nanofluids.

vi. **Application-specific Innovations**

Electronics and Data Centers

Innovations in nanofluid-based cooling systems will be driven by the growing need for effective cooling solutions in electronics and data centers.

Subsequent investigations will prioritize the development of nanofluids that possess enhanced thermal conductivity and dielectric characteristics. This will facilitate the effective cooling of high-power electronic equipment and reduce energy usage in data centers.

Automotive and Transportation

Nanofluids have the potential to significantly enhance the cooling system efficiency of internal combustion engines and electric vehicles, making them indispensable in the automobile sector. Future advancements will focus on refining the composition of nanofluids for use in automotive applications. This will involve ensuring that the nanofluids are compatible with existing cooling systems and improving the overall performance of vehicles.

Healthcare and Biotechnology

Nanofluids have considerable promise in the field of healthcare and biotechnology. One application of magnetic nanofluids is in hyperthermia cancer treatment, where they can be employed to specifically heat and eradicate cancer cells. Subsequent investigations will examine the utilization of nanofluids in precise drug administration, medical visualization, and additional biomedical uses, thus facilitating the development of groundbreaking healthcare remedies.

Energy Sector

The industrialization of nanofluids has the potential to greatly assist the energy sector. Nanofluids have the potential to increase the efficiency of cooling systems in nuclear reactors, hence enhancing safety and performance. Nanofluids can be utilized in oil and gas extraction to improve heat transmission within drilling fluids, hence enhancing the efficiency of extraction procedures. Future developments will prioritize the optimization of nanofluid formulations for various applications and guaranteeing their safe and efficient utilization.

vii. **Challenges and Future Directions**

Addressing Health and Safety Concerns

It is imperative to comprehensively investigate the potential health and safety hazards linked to the utilization of nanofluids. Subsequent investigations will prioritize the comprehension of the toxicity and environmental consequences of nanoparticles, as well as the formulation of secure protocols for their handling and disposal. To ensure the safe industrialization of nanofluids, it will be crucial to conduct thorough risk evaluations and establish clear safety rules.

Bridging the Gap between Research and Industry

In order to facilitate the transfer of nanofluids from research laboratories to industrial applications, it is essential to establish a connection between academia and industry. Collaborative research activities, business partnerships, and technology transfer programs will be crucial in expediting the commercialization of nanofluids. Our future endeavors will prioritize the cultivation of these partnerships and the establishment of platforms to facilitate the sharing of knowledge and transfer of technology.

10.6 SUMMARY

- The future prospects of nanofluids for manufacturing are highly encouraging, since they have substantial potential for breakthroughs in nanoparticle production, tailored formulations, environmental sustainability, and integration with developing technology.
- In order to achieve widespread implementation of nanofluids in different industries, it would be essential to tackle difficulties pertaining to stability, cost-efficiency, standardization, and safety.
- Nanofluids have the potential to significantly transform thermal management systems as research and development activities advance, providing novel solutions for the issues of the 21st century.
- Advancing the industrialization of nanofluids necessitates cooperation, interdisciplinary investigation, and a dedication to sustainability and safety. This will pave the way for a future in which nanofluids have a prominent role in improving efficiency and performance in various applications.

REFERENCES

1. Anctil, A., & Vasilis, F. (2012). Life cycle assessment of organic photovoltaics. *Third Generation Photovoltaics*. https://doi.org/10.5772/38977.
2. Malika, M., & Sonawane, S. S. (2024). Ecological optimization and LCA of TiO_2-SiC/ water hybrid nanofluid in a shell and tube heat exchanger by ANN. *Proceedings of the Institution of Mechanical Engineers, Part E: Journal of Process Mechanical Engineering, 238*(1), 45–55.
3. Malika, M., Jhadav, P. G., Parate, V. R., & Sonawane, S. S. (2023). Synthesis of magnetite nanoparticle from potato peel extract: its nanofluid applications and life cycle analysis. *Chemical Papers, 77*(2), 1081–1094.
4. Said, Z., Rahman, S. M. A., Assad, M. E. H., & Alami, A. H. (2019). Heat transfer enhancement and life cycle analysis of a Shell-and-Tube Heat Exchanger using stable CuO/water nanofluid. *Sustainable Energy Technologies and Assessments, 31*, 306–317.
5. Barberio, G., Scalbi, S., Buttol, P., Masoni, P., & Righi, S. (2014). Combining life cycle assessment and qualitative risk assessment: The case study of alumina nanofluid production. *Science of the Total Environment, 496*, 122–131.

Index

For Product Safety Concerns and Information please contact our EU
representative GPSR@taylorandfrancis.com
Taylor & Francis Verlag GmbH, Kaufingerstraße 24, 80331 München, Germany

www.ingramcontent.com/pod-product-compliance
Ingram Content Group UK Ltd.
Pitfield, Milton Keynes, MK11 3LW, UK
UKHW022314100726
473146UK00009B/474